VANCOUVER,
CITY ON THE EDGE

Living with a dynamic
geological landscape

Vancouver, City on the Edge

John Clague
&
Bob Turner

TRICOUNI PRESS
VANCOUVER

National Library of Canada Cataloguing in Publication Data

Clague, J. J. (John Joseph), 1946-
    Vancouver, city on the edge : living with a dynamic geological landscape / John Clague, Bob Turner ;
illustrations drawn or modified by Richard Franklin.

    Includes bibliographical references and index.
    ISBN 0-9697601-4-0

    1. Geology—British Columbia—Vancouver. 2. Geology—British Columbia—Lower Mainland.
I. Turner, Bob, 1954- II. Franklin, Richard, 1956- III. Title.
QE187.C52 2003   557.11´33   C2003-910357-9

Available from:
    Tricouni Press Ltd.
    3649 West 18th Avenue
    Vancouver, BC, Canada V6S 1B3
    Phone and fax: 604-224-1178
    Email: books@tricounipress.com

and from:
    Gordon Soules Book Publishers Ltd.
    1359 Ambleside Lane
    West Vancouver, BC, Canada V7T 2Y9
    Phone: 604-922-6588
    Fax: 604-688-5442
    Email: books@gordonsoules.com

Edited, designed, and typeset by Glenn and Joy Woodsworth
Set in Minion and Charlotte Sans
Printed and bound in Canada by Friesens
Printed on acid-free paper ∝

The authors and publisher gratefully acknowledge the support of
the Canadian Geological Foundation and
the Geological Survey of Canada.

In memory of William H. Mathews (1919–2003), who for over thirty years was a professor in the Department of Geological Sciences at the University of British Columbia. Bill was a legend in Canadian geology. He made important contributions in many fields of earth science and mentored many graduate students, including one of us (Clague), who was strongly influenced by him.

Field trip stops

The following stops, set off with a green background, illustrate many of the features discussed in this book.

Contents

Landscapes of the Lower Mainland.

Centre: satellite image of the Fraser River delta, Point Grey, and Tsawwassen.

Top left: Siwash Rock in Stanley Park. Top middle: downtown Vancouver, view north from near Cambie Street.

Top right: Garibaldi Lake north of Squamish.

Bottom right: Fraser River east of Mission. Bottom middle: prime agricultural land on the Fraser River delta.

Bottom left: Mount Baker as seen from Vancouver.

Our magnificent landscape

Vancouver is considered by many people to be the most beautiful city in North America. For this we can thank Mother Nature. We are blessed with a spectacular setting. To the west is the Strait of Georgia, an inland sea, separated from the Pacific Ocean by Vancouver Island; to the north and southeast are the high peaks of the Coast and Cascade Mountains. Vancouver itself lies within the Fraser Lowland or Fraser Valley, a triangular-shaped, low-lying area across which B.C.'s largest river flows on the final leg of its long journey to the sea.

This book tells the story of Vancouver's geological landscape, or "geoscape," a story that began in the age of dinosaurs, eons before humans appeared on Earth. It explains the geology of the Lower Mainland, answering such questions as "What are the rocks that underlie Vancouver?" and "How did they form?" The book also describes the natural Earth processes that have sculpted our landscape since time immemorial and continue to do so today. Local mountains were born of colliding crustal **plates** [note: **bold** words and phrases are defined in the glossary on page 175], earthquakes, and volcanic eruptions. Glaciers and streams have carved deep valleys and deposited their **sediment** load in lowlands, lakes, and the sea. Landslides displace rock and soil, often catastrophically. Waters dissolve, transport, and deposit metals and minerals. Some of these natural processes pose potential hazards to people and property. Large earthquakes and floods, for example, can not only

Highview Lookout

On a clear day, this viewpoint, which is located along the road to the Cypress Bowl ski area, provides a magnificent view of Vancouver, the lower Fraser Valley, Mount Baker, Burrard Inlet, and the southern Strait of Georgia. In the background, from west to

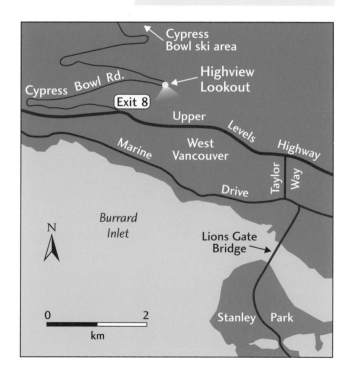

east, are Vancouver Island, the southern Gulf Islands, the San Juan Islands, and the Cascade Mountains, dominated by Mount Baker volcano. Note the three major elements of the landscape — mountains, **Ice Age** uplands, and lowlands crossed by the Fraser River and its tributaries. The Fraser River and its delta are visible in the middle distance. Also note the asymmetrical profiles of Burnaby Mountain, Capitol Hill, and Stanley Park, which result from the southerly **dip** of the layered **Cretaceous** and **Tertiary sedimentary rocks** that core them (pages 25, 28).

Take a short walk down the road from the Highview Lookout to see a good exposure of glacial **till**, stony sediments left by the ice sheet that covered the Lower Mainland 15,000 years ago (but please watch for traffic!). Look for a small stream about 200 m down the road on the right side. The stream cascades over the till just above road level. Excellent outcrops of till can also be seen along Cypress Creek just east of the road leading to the downhill ski area in Cypress Bowl. The best exposures are 700 m beyond the turnoff to the cross-country ski area.

cause billions of dollars of property damage but also injure and kill people. Such catastrophes are uncommon events in the time scale familiar to humans but they are frequent and inevitable on the longer, geological time scale.

Our goal in this book is to provide you with a better understanding of our local geology and its value to society. The resources and land needs of our rapidly growing population demand decisions as to where and how we accommodate growth. By understanding our geological landscape, we promote its wise use: protecting vital resources, avoiding hazards, and reducing the risk of hazards that cannot be avoided. It is here that geoscientists make their greatest contribution. The quality of life in the Lower Mainland will partly depend on how well we incorporate geological knowledge into our land-use decisions.

The facing map shows the distribution of the three major landscape elements of the Lower Mainland — mountains, Ice Age uplands, and modern lowlands. Mountain areas (Coast and Cascade Mountains) comprise rugged bedrock ridges and peaks and intervening steep-walled valleys.

The other two landscape elements are within the Fraser Lowland. Higher parts of the Fraser Lowland are gently rolling surfaces, ranging from about 15 to 250 m above sea level and underlain by thick Ice Age sediments. Flat lowlands occur along the Fraser River and its tributaries and are underlain by modern sediments.

Landscape elements —
three-part harmony

VANCOUVER residents can rightly take pride in the physical setting of their city, but most of us know little about the origin of this setting. Let's start our tour of the geology of Vancouver by looking a little more closely at the local landscape.

The landscape of the Lower Mainland consists of three **physiographic** elements or domains — mountains, Ice Age uplands, and modern lowlands. The mountain domain includes the Coast and Cascade Mountains, rugged areas of steep slopes, snow- and glacier-clad peaks, and deep valleys. Small outliers of the mountain domain rise up above low-lying areas in the Fraser Lowland — Burnaby and Sumas Mountains are examples. Our mountains are

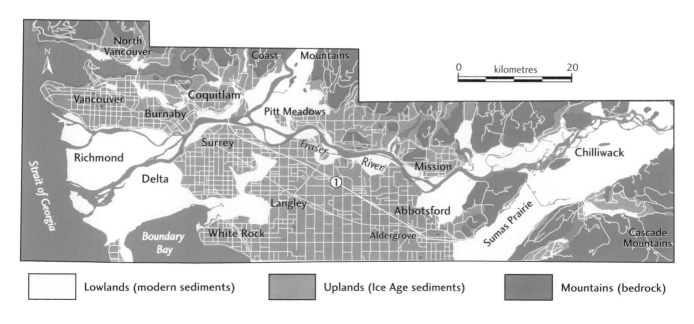

| | Lowlands (modern sediments) | | Uplands (Ice Age sediments) | | Mountains (bedrock) |

formed of a wide variety of rocks that are as much as 200 million years old. The upland domain, the second main physiographic element, includes much of the Fraser Valley and is the place where most of us live. It is a rolling surface ranging in elevation from about 10 to 200 m above sea level and is underlain mostly by sediments that date to the Ice Age, 2 million to 10,000 years ago. The youngest, and lowest, of the three domains is the modern lowland domain. Lowlands are less than 10 m above sea level and have formed since the Ice Age, within the last 10,000 years. The lowland domain includes the **floodplain** of the Fraser River and its tributaries, the Fraser and Squamish river **deltas**, and the shorelines of the Strait of Georgia.

The following several sections of this book are based on this three-fold subdivision of the landscape. In the next chapter, we examine the rocks of Vancouver, which are found principally in the mountains but also occur in the upland domain. This chapter is followed by one on the Ice Age, with a discussion of the upland domain and the sediments that form it. Next is a chapter on modern lowlands, with an emphasis on the Fraser River and its delta.

The three major landscape elements of the Lower Mainland are visible in this photo taken at the south end of the Port Mann Bridge — the floodplain of the Fraser River in the foreground, Ice Age upland on which Coquitlam is located, and the Coast Mountains in the background.

This simplied geological map shows the distribution of the main types of Earth materials at and beneath the surface in the Lower Mainland. The Fraser Lowland is the site of an ancient trough in which there are thousands of metres of Cretaceous and Tertiary sandstone, mudstone, and conglomerate. Ice Age sediments overlie these sedimentary rocks and form much of the surface of the lowland. Rivers, primarily the Fraser and its tributaries, have deposited their sediment loads in valleys in the Ice Age sediments. Today, clay, silt, and sand from the Fraser River continue to accumulate in the Strait of Georgia.

13

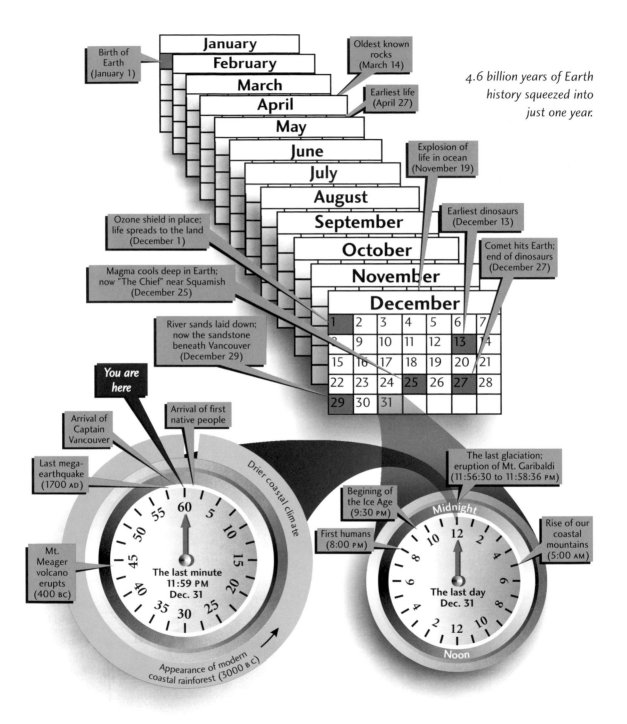

4.6 billion years of Earth history squeezed into just one year.

Birth of Earth (January 1)

Oldest known rocks (March 14)

Earliest life (April 27)

Explosion of life in ocean (November 19)

Ozone shield in place; life spreads to the land (December 1)

Earliest dinosaurs (December 13)

Comet hits Earth; end of dinosaurs (December 27)

Magma cools deep in Earth; now "The Chief" near Squamish (December 25)

River sands laid down; now the sandstone beneath Vancouver (December 29)

January February March April May June July August September October November December

1	2	3	4	5	6	7
8	9	10	11	12	13	14
15	16	17	18	19	20	21
22	23	24	25	26	27	28
29	30	31				

You are here

Arrival of first native people

Arrival of Captain Vancouver

Last mega-earthquake (1700 AD)

The last glaciation; eruption of Mt. Garibaldi (11:56:30 to 11:58:36 PM)

Drier coastal climate

Begining of the Ice Age (9:30 PM)

First humans (8:00 PM)

Rise of our coastal mountains (5:00 AM)

Mt. Meager volcano erupts (400 BC)

The last minute 11:59 PM Dec. 31

Midnight

The last day Dec. 31

Noon

Appearance of modern coastal rainforest (3000 BC)

Deep time —
the record written in the rocks

THE STORY of how Vancouver's spectacular landscape came to be is a geological one, written in the rocks beneath our feet and in the immensity of geological time. This is a fantastic story, as amazing as any that a Hollywood scriptwriter could imagine. Let's take a walk through time and see what the rocks in the mountains around us tell us about the origin of our home place. Before we begin our journey, however, some words about geological, or "deep," time.

Geological time

Geologists estimate the Earth to be 4.6 billion years old — that's 4,600,000,000 years! This length of time is incomprehensible to most of us. To give you a sense of what deep time means, imagine that you could squeeze all 4.6 billion years into a single calendar year beginning one second after midnight on January 1. The earliest life would originate on April 27 of our imaginary year. The earliest dinosaurs appeared on December 13. It was not until 8 PM December 31 that humans evolved from more primitive primates. The Ice Age ended at 11:59 PM December 31 (in real time 10,000 years ago). All recorded history falls within the last 30 seconds. Captain George Vancouver arrived less than a second before the end of the year (AD 1794 in real time)! Eighty years, which is the average length of a person's life in Canada, corresponds to a small fraction of one second on our imaginary clock. You get the idea: geological time is vast, so vast as to be beyond comprehension.

Why is this important? Because, with enough time, natural processes such as river erosion, landslides, volcanic activity, and glaciation will profoundly modify the Earth. In the vast expanse of time, mountain ranges appear and disappear, ice sheets come and go, new species of plants and animals arise, and others become extinct. The present landscape and rocks of Vancouver are the cumulative product of the last 200 million years of Earth history.

Colliding plates and exotic terranes

On a time scale of years or even centuries, the Earth appears essentially static and unchanging. Nevertheless, looks can be deceiving, for on the longer time scale, that of geological time, the Earth is dynamic and undergoes profound change. The landscape of British Columbia today, for example, bears little resemblance to that of 50 million years ago. What caused our landscape to change so much?

The outer shell, or **crust**, of the Earth is about 5 to 250 km thick

The major tectonic plates that make up the crust of the Earth. The plates are slowly moving with respect to one another on top of the mantle. Vancouver is near the boundary of the westward-moving North American Plate and the smaller Juan de Fuca Plate. See page 134 for a more detailed map of western North America

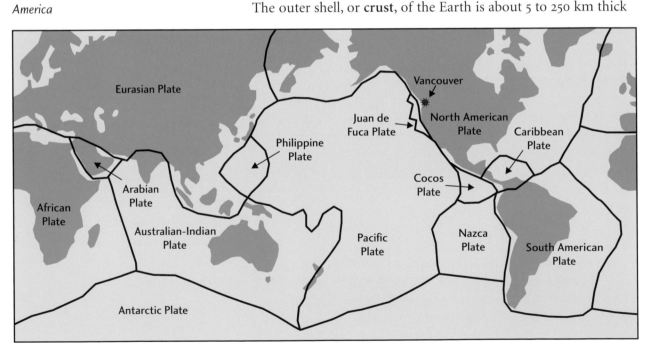

and consists of a mosaic of seven large plates and a number of smaller ones that float independently of one another on the Earth's **mantle**. The six large plates, including, for example, the plate that forms North America and part of the Atlantic Ocean basin (North American Plate), are far bigger than the largest continents. Other plates are much smaller, for example, the Juan de Fuca Plate, which underlies the floor of the east Pacific Ocean between northern California and southern B.C.

Crustal plates move because the mantle on which they rest is plastic, somewhat akin to toffee or silly putty. Plumes of super-heated mantle rock rise from deep within the Earth, move laterally approximately parallel to the Earth's surface, becoming cooler in the process, and then sink back down towards the Earth's **core** (next page). This process, termed **convection**, is driven by heat produced through the decay of radioactive elements in the Earth. Convection itself drives the slow, inexorable movement of plates at the Earth's surface, a process that geologists call **plate tectonics**.

About 180 million years ago, the North American Plate began to separate from Europe along a rift that eventually developed into the Atlantic Ocean. Since then, the North American Plate has moved slowly but steadily westward, colliding with and overriding oceanic plates composed mainly of **basalt**. This collision process, termed **subduction**, is responsible for compression and thickening of crustal rocks, the rise of mountains, volcanic eruptions, and earthquakes. Island chains, large submarine volcanoes (**seamounts**), and mud deposits on the ocean floor were slowly scraped off the downgoing oceanic plates at the edge of the continent. In this way, far-travelled crustal fragments, or **terranes**, collided with and became attached to North America. Geologists have shown that all of B.C., with the exception of the Rocky Mountains and the Interior Plains of northeastern B.C., consists of a complex mosaic of terranes that successively collided with, and became part of, the growing North American continent. Each terrane comprises a unique assemblage of rocks and each is separated from neigh-bouring terranes by **faults**. Some of the terranes originated in equatorial areas and travelled thousands of kilometres before reaching their current resting places. As they docked with North

Living on the edge — a slice into the Earth illustrating the titanic collision that forces oceanic crust beneath the continental crust of western North America and into the hotter regions of the Earth. This process leads to earthquakes, volcanic eruptions, and mountain building, which are the hallmarks of Canada's Pacific margin.

Holy subduction!

Over 25,000 km of Pacific Ocean crust, enough to extend more than halfway around the Earth, have been subducted beneath western North America over the past 180 million years. Subduction is responsible for the volcanoes and mountain ranges of British Columbia; its history is preserved in the rocks that underlie Vancouver today.

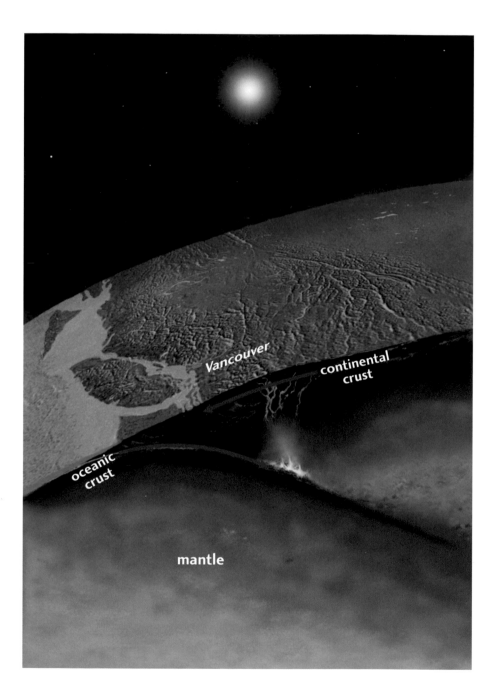

America, the terranes were squeezed, folded, and split apart along faults, forming mountain ranges. They were also eroded by rivers, glaciers, and gravity, all of which work to lower the land.

These processes are still active today. A fundamental principle used by geologists to reconstruct past events is the Principle of Uniformitarianism, loosely paraphrased as "the present is the key to the past." In other words, the physical processes that presently are at play on Earth are the same as those that have shaped it during the past. You may find it hard to believe that a landscape can be significantly altered by the processes described above, given the snail-like pace at which they operate. Remember, however, that these processes have literally all the time in the world. A plate moving at 4 cm per year may not move far in a human lifetime (about 3 m), but it would move 4 km in one million years, and one million years is "small change" in the history of the Earth. Given enough time, mountains can form or disappear, large valleys can be cut, and rivers can change their course or even their direction of flow.

Rocks that formed in the Earth's pressure cooker

Erosion of the Coast and Cascade Mountains has exposed **granitic** and **metamorphic rocks** that formed deep in the Earth's crust more than 100 million years ago. A trip by car from Vancouver to Squamish traverses these rocks — **granites** of the North Shore mountains and Stawamus Chief, and ancient metamorphosed sedimentary and **volcanic rocks** at Britannia Beach.

The oldest rocks in the Vancouver area occur at the surface in the Coast Mountains north of Vancouver, beneath the Fraser Lowland at depths up to 10 km or more, and in the Cascade Mountains south of Chilliwack. They are what geologists call metamorphic rocks — rocks that were originally sedimentary or **igneous** in origin, for example **mudstone** or basalt, but that have been subjected to high temperatures and pressures deep within the Earth.

British Columbia's southwest coast — "action central"

The southwest coast of British Columbia lies at the boundary of two great crustal plates — the oceanic Juan de Fuca Plate, which underlies the easternmost North Pacific Ocean from northern California to central Vancouver Island, and the much larger, continental North American Plate. The two plates are converging at a rate of about 4 cm per year. They meet head-on about 200 km off the coast; there the Juan de Fuca Plate begins its slow descent beneath western B.C.

Collision of the two plates and subduction of one beneath the other cause earthquakes on the west coast and are also responsible for the chain of active volcanoes at our doorstep. They are also creating and modifying our landscape. Detailed surveys indicate that the Coast Mountains are rising at rates of up to a few millimetres per year. The uplift prevents the rugged mountains from being levelled in the face of continuous erosion by streams, glaciers, and landslides. Our landscape is active — the earth movements that created it continue today!

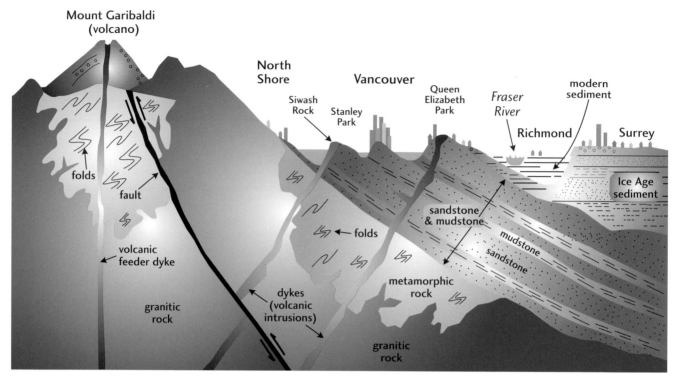

Mount Garibaldi
(volcano)

folds

fault

volcanic
feeder dyke

granitic
rock

North
Shore

Vancouver

Siwash
Rock

Stanley
Park

Queen
Elizabeth
Park

*Fraser
River*

modern
sediment

Richmond

Surrey

Ice Age
sediment

sandstone
& mudstone

mudstone

sandstone

folds

metamorphic
rock

dykes
(volcanic
intrusions)

granitic
rock

Diagrammatic, north-south cross-section of the upper part of the Earth showing the general geometry and contact relationships of major rock and sediment types that underlie the Vancouver area. The vertical scale is exaggerated. The oldest rocks are metamorphic (green) and granitic (purple) rocks that form the local mountains and extend beneath the Fraser Lowland. Sandstone and mudstone (brown) on which much of Vancouver is built were deposited on top of these old rocks and were later tilted to the south (right). Ice Age sediments (orange) were laid down on all older rocks. The youngest geological units in the area are the volcanic rocks of Mount Garibaldi (red) and modern sediments of the Fraser River delta (yellow).

We can better understand how these ancient rocks were created by considering our own **subduction zone.** The Earth gets hotter with depth. Data from deep mines and drill holes worldwide indicate that the average temperature rise is about 5 to 75° C per kilometre depth. At Vancouver, crustal rocks 15 to 30 km below the surface (30 km is the distance from downtown Vancouver to Tsawwassen) are estimated to have temperatures of 300 to 400° C. At these depths, the rocks are subject to great pressure due in part to the weight of the overlying rock, and they behave more like putty than solid rock. Under such conditions, sedimentary and

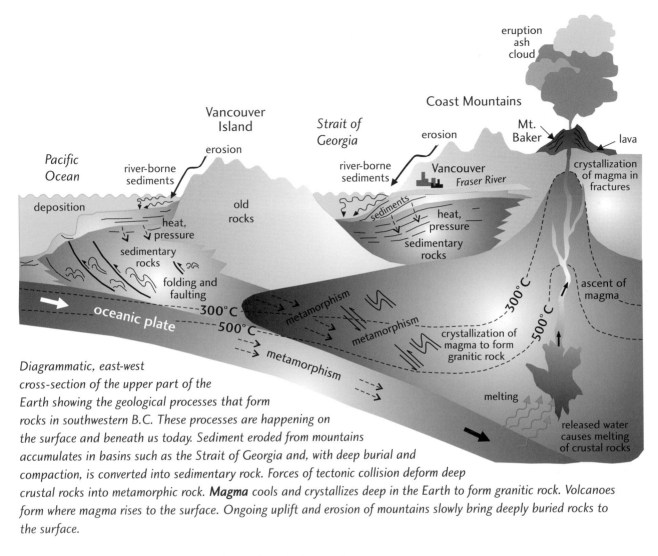

eruption
ash
cloud

Coast Mountains

Vancouver
Island

*Strait of
Georgia*

Mt.
Baker

lava

*Pacific
Ocean*

erosion

erosion

river-borne
sediments

river-borne
sediments

Vancouver

crystallization
of magma in
fractures

deposition

Fraser River

sediments

heat,
pressure

heat,
pressure

old
rocks

sedimentary
rocks

sedimentary
rocks

ascent of
magma

folding and
faulting

metamorphism

300°C

300°C

oceanic plate

500°C

metamorphism

metamorphism

crystallization of
magma to form
granitic rock

500°C

metamorphism

melting

released water
causes melting
of crustal rocks

*Diagrammatic, east-west
cross-section of the upper part of the
Earth showing the geological processes that form
rocks in southwestern B.C. These processes are happening on
the surface and beneath us today. Sediment eroded from mountains
accumulates in basins such as the Strait of Georgia and, with deep burial and
compaction, is converted into sedimentary rock. Forces of tectonic collision deform deep
crustal rocks into metamorphic rock.* **Magma** *cools and crystallizes deep in the Earth to form granitic rock. Volcanoes
form where magma rises to the surface. Ongoing uplift and erosion of mountains slowly bring deeply buried rocks to
the surface.*

volcanic rocks undergo mineral, physical, and chemical changes,
and slowly become metamorphic rock. Shearing forces created by
the Juan de Fuca Plate sliding past the base of the overlying North
American Plate cause the metamorphic minerals to align as they
grow, giving many of these rocks a conspicuous layering, or
foliation. Other rocks have been heated but not deformed and
thus lack foliation.

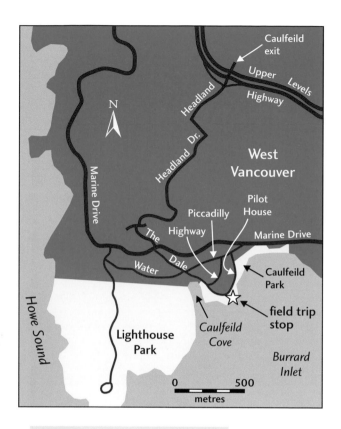

Gneiss (dark, banded rock on which the person is standing) cut by granite (light-coloured rock) at Caulfeild Cove. When the gneiss was deep in the crust and very hot, magma was injected along fractures within it. The magma cooled and crystallized to form the granite. The granite contains dark blocks "plucked" from the gneiss.

Caulfeild Park — the deep Earth exposed

Caulfeild Park is a "must see" for geology students young and old. As you walk along the very scenic shore in the park, you will cross some of the oldest rocks in the Vancouver area. These rocks formed deep within the Earth under very high temperature and pressure. Since then, they have been elevated tens of kilometres

All right, you say, but if the metamorphic rocks were created at great depth in the Earth's crust, why are they found today at the surface? Here's the answer — the metamorphic rocks have been slowly uplifted from the depths at which they formed by mountain-building processes. As the Coast and Cascade Mountains rose over a period of tens of millions of years, erosion stripped kilometres of rock from the mountains, eventually exposing their roots. Imagine the forces needed to elevate so vast an area of the crust!

by Earth forces, and the overlying rocks have gradually eroded away. There are two main types of rocks at Caulfeild Park — light-coloured granitic rocks consisting mainly of crystals of the minerals **quartz** and **feldspar**, and dark banded metamorphic **gneiss** (the "Caulfeild Gneiss"), which contains abundant feldspar and black **hornblende**. The gneiss formed through recrystallization under high temperature and pressure from pre-existing sedimentary or volcanic rock. It is cut by numerous **dykes** of granitic rock of different ages, which suggests that granitic magma filled fractures in the gneiss and, therefore, that the gneiss is older than the dykes. But wait a minute, look more closely: some of the granitic rocks seem to grade into the gneiss, without a sharp boundary. Could some of the granitic rocks have formed directly from the gneiss without actually having melted? Some geologists think so. Examine these rocks closely and see what you think. You can use the geological principle of **crosscutting** relationships to sort out the relative ages of the various dykes and the gneiss. This is a fun geological "brain teaser"!

*Granite, a common rock in the Coast Mountains of B.C., with its characteristic "salt and pepper" appearance. The grey, translucent-looking mineral is quartz (Q); the white to slightly pinkish mineral is feldspar (F); and the black mineral is **biotite**, a mica. The field of view is about 10 cm across.*

The Coast Mountains — granite country

The Coast Mountains extend from Vancouver's North Shore nearly 1000 km to the Yukon and Alaska. This range is one of world's great mountain belts, and much of it consists of granitic rock. Granitic rock is formed mainly of interlocking crystals of feldspar and quartz, the two most common minerals in the Earth's crust. The feldspar and quartz grains, along with lesser amounts of hornblende, biotite mica, and other minerals, crystallized from molten rock, or magma, as it slowly cooled deep within the Earth's crust. Magma forms in the lower part of the crust due to the release of water and other fluids from the underlying subducting oceanic crust as it descends, becomes hot, and undergoes mineral transformation. Complete melting occurs at temperatures above 1000° C. The great belt of Coast Mountains granitic rock thus records a story of ancient subduction and magma generation. It also represents the deepest part of the magma "plumbing system" that fed volcanoes at the surface and is the eroded root of ancient volcanic chains.

The granitic mass of Stawamus Chief rises over 600 m above Highway 99 at Squamish. Its great, near-vertical, glacially polished faces are a legacy of the Ice Age. The granite crystallized deep in the crust 100 million years ago.

Geologists recognize many different types of granitic rocks, depending on their mineral and chemical composition. Technically, most of the granitic rocks in the Coast Mountains are **granodiorite** and **tonalite**; true granite is less abundant. But outside the circle of geologists, "granite" is commonly applied to all granitic rocks, and the term is used in this sense throughout this book. In this context, the Coast Mountains are "granite country."

Most of the granites in the Vancouver area formed between 150 and 80 million years ago, but some are older and a few are less than 20 million years old. Like metamorphic rocks, they have been raised to the surface while the overlying cover rocks have been eroded away.

Vancouver 70 million years ago —
a tropical paradise

The rocks that form the foundation of downtown Vancouver and most of the Fraser Lowland are **sandstone**, mudstone, and **conglomerate** that formed about 70 to 40 million years ago, during and after the time of the dinosaurs. Geologists have placed these sedimentary rocks in two units, the Cretaceous-age "Nanaimo Group" and the "Huntingdon Formation" of **Paleocene** and **Eocene** age. They overlie and thus are younger than the metamorphic and granitic rocks described above and were deposited by rivers flowing off ancient coastal mountains. They have not been subjected to the high temperatures and pressures of the older rocks because they were never buried deep in the Earth.

Sandstone is well exposed along the Stanley Park seawall from Third Beach to Prospect Point. It is also present along the shore of Burrard Inlet west of the Vancouver Planetarium to near Jericho Beach; it is found in building excavations in downtown Vancouver; and it underlies Capitol Hill and Burnaby Mountain. Mudstone of the same age underlies the floor of Burrard Inlet. Mudstone is more vulnerable to erosion than sandstone and consequently has been more deeply eroded.

The striking asymmetry of the surface profiles of Stanley Park, Capitol Hill, and Burnaby Mountain are controlled by the inclination of the layered sedimentary rocks that underlie them. These sedimentary rocks consist of layers, or **beds**, that were nearly flat at the time they were deposited but were tipped as the Coast Mountains rose relative to the Fraser Lowland. Today, these rocks slope, or dip down, about 10 degrees to the south, and land surfaces associated with them are also inclined in this direction.

Layered sandstone exposed in the tidal zone at Ferguson Point in Stanley Park. Wave erosion of layers of different hardness has created the ribbed texture on the rock surface. Originally horizontal, the sandstone was tilted gently to the south during uplift of the North Shore mountains.

Stanley Park — a stroll through geological time

You can take a nice stroll through geological time along the Stanley Park seawall between Prospect Point and Second Beach. The rocks that you see along the seawall are mainly sandstone of Cretaceous, Paleocene, and Eocene age. The sandstone beds dip gently to the south, which is what gives Stanley Park its asymmetric profile. When you drive along the Stanley Park causeway towards Vancouver, you are descending the gentle southerly slope of the sedimentary rocks that underlie the park. Because the rocks are inclined to the south, they become younger as one walks along the seawall towards Third Beach. Thus, at Prospect Point, the rocks are late Cretaceous in age, about 70 million years old. In contrast, just south of Third Beach, the rocks are Tertiary in age, dating to some time between about 60 and 50 million years ago. So, as you stroll the 2 km from Prospect Point to Second Beach, you walk through millions of years of time!

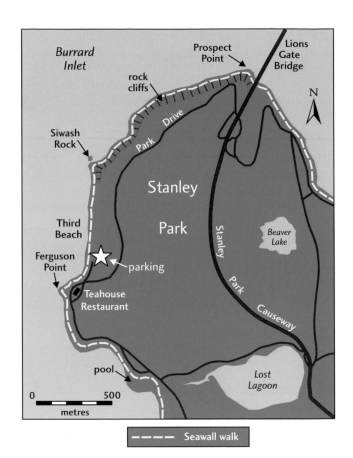

Similar sedimentary rocks form most of the Gulf Islands and probably underlie much of the Strait of Georgia and the low country south of Bellingham, Washington. They also extend under much of the Fraser Lowland as a layered sequence up to several kilometres thick. However, these rocks are uncommon at the surface as they are deeply buried beneath much younger sediments left by glaciers during the Ice Age and by rivers after the glaciers had disappeared.

Siwash Rock, a Vancouver landmark located in Stanley Park. This pillar of resistant volcanic rock is part of a dyke (see page 20) that was intruded into the sandstone (right side of photo) about 32 million years ago. It was once a rocky point on the coastal cliff but became separated from it by wave erosion and weathering.

The sandstone was deposited by ancient rivers and streams that carried their sandy load from a nearby mountain range to the sea. You might wonder how geologists know these rocks are 50 to 70 million years old. The rocks contain fossilized leaves and stems of plants that are diagnostic of the Cretaceous and Tertiary periods, as well as tiny pollen grains and spores.

At Prospect Point and Siwash Rock, the Cretaceous sandstone is cut by black, basaltic dykes or **sills**. The basalt is more resistant to erosion than the sandstone and thus forms rugged cliffs. Siwash Rock stands as an erosional remnant, Vancouver's only example of a **sea stack**. Take a closer look at the rocks next to the seawall at Siwash Rock. You can see that the sandstone has been "baked" by injection of the basaltic dykes along fractures. Geologists have dated the basalt at Prospect Point using the **potassium-argon dating** method to 32 million years old. It is thus much younger than the sandstone into which it has been injected.

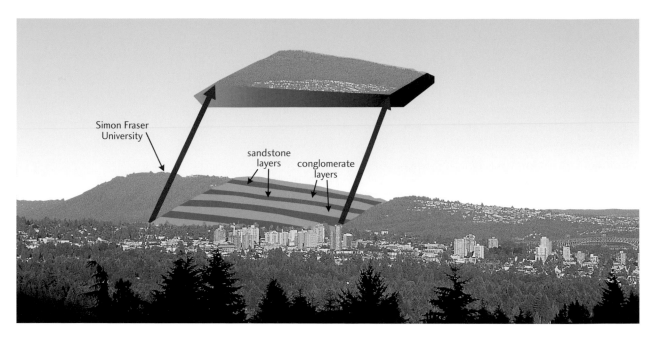

Burnaby Mountain as seen from the northwest, with an imaginary cut-away exposing the bedrock beneath the mountain. The cut-away shows schematically the tilted sandstone and conglomerate layers that underlie the mountain and are responsible for its pronounced surface asymmetry.

Fossil fronds of a Cretaceous palm, Phoenicites imperialis, found at the site of Malaspina College in Nanaimo. The field of view is about 30 cm across.

The sandstone and mudstone contain **fossils**, traces of black, carbonized plant life that were buried in the sediments. The fossils tell us a lot about what things were like at Vancouver 70 to 50 million years ago. And what a different place it was! The vegetation was tropical and lush, with huge ferns, palms, and exotic trees that are now long extinct. Dinosaurs roamed the Earth during the early part of this interval, although no dinosaur fossils have been found in the rocks at Vancouver. The days of the dinosaurs, however, were numbered: these amazing animals disappeared 65 million years ago, possibly as a result of an asteroid impact in the Gulf of Mexico off Yucatan. What is now Vancouver was a lowland or estuary drained by rivers that deposited their loads of **gravel**, **sand**, **silt**, and **clay**. Large swamps dotted the landscape. Decaying plant material accumulated in some of these swamps and was later transformed into **coal**. Seams of coal beneath Nanaimo and Nanaimo Harbour were mined from about 1852 until 1953.

About 30 million years ago, long after they had formed, the sedimentary rocks were intruded by basaltic magma. Basaltic magma comes from deeper in the Earth than granitic magma, from depths of perhaps 50 km in the mantle. It is hotter and more fluid than granitic magma and is able to rise to the surface, or near it, before cooling and turning to rock. Basalt is a fine-grained, dark coloured, commonly black rock. In the Vancouver area, basalt occurs as dykes, sills, and **lava flows**. Dykes are sheet-like volcanic intrusions that cut older rocks; a dyke forms much of Siwash Rock in Stanley Park. Sills are intrusions that are parallel to the layers in sedimentary or metamorphic rocks. Basalt at Grant Hill on the north side of the Fraser River east of Haney is a good example of a sill. Lava flows are produced when magma erupts onto the ground and flows downhill. You can see good examples of lava flows at Brandywine Falls north of Squamish (page 144).

Basalt at Queen Elizabeth Park in Vancouver was quarried in the early decades of the last century for use as road-bed material. The quarry was later transformed into the spectacular gardens of Queen Elizabeth Park, which attract hundreds of thousands of visitors each year. Very few of our visitors are aware that these gardens were formerly a rock quarry, the scene of heavy industrial activity.

Fossils and evolution

William Smith, one of the founders of the modern science of geology, wrote the following in his notes in 1796, more than 60 years before publication of Charles Darwin's book *Origin of Species:*

> Fossils have long been studied as great curiosities, collected with great pains, treasured with great care and at a great expense, and showed and admired with as much pleasure as a child's rattle or a hobby-horse is shown and admired by himself and his playfellows, because it is pretty; and this has been done by *thousands who have never paid the least regard to that* wonderful order and regularity with which Nature has disposed of these singular productions, and assigned to each class its particular stratum. (Simon Winchester, 2001, *The Map That Changed the World,* HarperCollins, p. 119–120.)

Smith, a canal engineer in England at the close of the nineteenth century, made a startling discovery that was to turn the fledgling science of geology and a central plank of established Christian religion on their heads. He saw that the rocks he was excavating were arranged in layers and that the fossils present in one layer were different from those in adjacent layers. The specimens of one kind of fossil, say corals, were the same throughout one layer, but were subtly different from the corals in another bed. A period of time thus had elapsed between deposition of the two beds, and the same period of time spanned the existence of the two kinds of corals. What Smith had observed, and what many others since have demonstrated in far more detail, is the foundation for what today we call "evolution."

Evolution is the theory of the origin and perpetuation of new species of animals and plants. Its tenets are that each species has genetic variation that can be transmitted from generation to generation, that natural selection favours the survival of some of these variations over others, that new species have arisen and may continue to arise by these processes, and that widely divergent groups of animals and plants have arisen from the same ancestors. Ever since publication of *Origin of Species,* biological evolution has been heatedly debated. The theory is accepted today by most scientists, including nearly all geologists and biologists, although there is much discussion about exactly how new species appear and disappear. Some people, however, reject evolution because they believe that the Earth is far too young for natural selection to play out. These people accept the edict from Archbishop James Ussher (1581–1656), during the Elizabethan era, that the Earth was created by God in October 4004 BC.

We won't get involved here in a discussion about the theory of evolution. It is important to note, however, that sedimentary rocks around the world have distinctive assemblages of fossils that change vertically through the rock succession, that is, through time. Dinosaurs are found in **Mesozoic** rocks but not in younger Tertiary rocks, and these reptilian creatures show systematic changes in species and form through the Mesozoic. The types of fossils that we find in sedimentary rocks also depend on the origin of the rocks. Some rocks were deposited in the sea and thus contain fossil corals, molluscs, fish, and other marine organisms. Other rocks were deposited on land, perhaps in swamps, river valleys, and lakes. Their fossils include land plants, vertebrate bones, and freshwater fish. Much can be inferred about the age and origin of a particular rock unit from the fossils it contains. A sequence of sandstone and mudstone containing dinosaur bones and fossil ferns probably formed in a swamp bordering a river, whereas a **limestone** containing corals is evidence of a reef in a shallow sea.

Most igneous and metamorphic rocks in southwestern B.C. do not

The ammonite Pachydiscus suciaenis *recovered from Cretaceous rocks (Nanaimo Group) on Hornby Island. The fossil is 10 cm in diameter. Ammonites are an extinct order of cephalopods, a group of marine animals that includes the nautiloids (e.g., pearly nautilus), cuttlefish, squid, and octopi.*

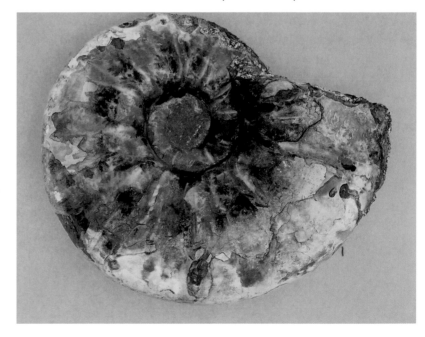

Imprints of fossil plants found in Cretaceous rocks on eastern Vancouver Island.
Top: Leaf of the angiosperm Viburnum; *the leaf is about 7 cm long.*
Bottom: A bouquet of the fern Asplenium *(left) and a frond of* Coniopteris *(right, about 8 cm long).*

contain fossils. Igneous rocks crystallize from magma, which contains no life. Fossils in deeply buried sedimentary rocks commonly were destroyed by high temperature and pressure, although some fossils may survive limited "cooking." For example, imprints (**moulds**) and remains of long-extinct marine molluscs can be found in weakly metamorphosed mudstone near Harrison Lake and in sandstone and mudstone on eastern Vancouver Island and several of the Gulf Islands.

Cretaceous and Tertiary sedimentary rocks exposed at Vancouver, in the Bellingham area, on eastern Vancouver Island, and on the Gulf Islands contain fossils of plants that lived tens of millions of years ago. The fossils include stems, leaves, pollen, and spores of a variety of land plants that grew in muggy tropical swamps and on wet floodplains. Imagine the swamps of Louisiana or Florida, but with a much more exotic vegetation, something like that in *Jurassic Park*, and you will have an idea what it was like at Vancouver 70 million years ago. Some of the plant remains accumulated to great thickness in swamps and eventually were buried beneath mud and sand washed into the swamps by nearby rivers. Subjected to pressure and heat, the buried plant material was transformed into coal, which is preserved as seams in Cretaceous rocks on eastern Vancouver Island, for example at Nanaimo and west of Campbell River. Small amounts of coal are also present in Cretaceous and early Tertiary sedimentary rocks around the Lower Mainland, including the Coal Harbour area in Stanley Park. Moulds and carbonized remains of fossil plants are still visible in coal seams and in beds of mudstone associated with the coal.

Volcanoes at our doorstep

The last geological chapter written in the rocks of Vancouver dates to about 3 million years ago, when volcanoes began to erupt along a line extending from southern B.C. to northern California. Since then, thousands of eruptions have produced the large conical volcanoes that collectively constitute the Cascade volcanic belt (see figure on page 134). This belt includes such well known volcanoes as Mount St. Helens, Mount Rainier, and Mount Hood. It also includes two volcanoes right at our doorstep, Mount Baker and Mount Garibaldi. The story of these two volcanoes is told elsewhere in this book (pages 137 and 146). Suffice it to say here that Mount Baker is a sleeping giant — an active and potentially dangerous volcano. Mount Garibaldi, on the other hand, has not erupted in over 10,000 years. Geologists hesitate to dismiss Mount Garibaldi as a "dead" volcano, but the likelihood of it erupting again in the next few hundred years is very low.

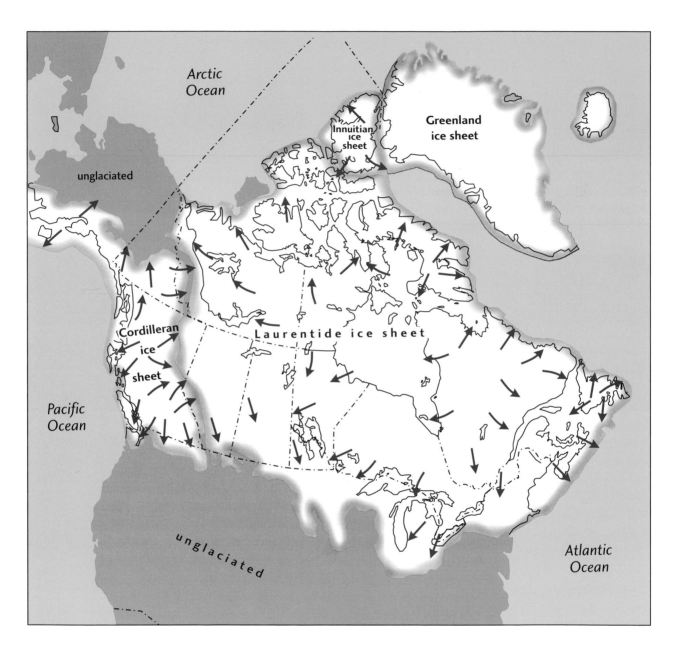

The maximum extent of the great North American ice sheets during the Ice Age. British Columbia was covered by the Cordilleran ice sheet, which bordered the much larger Laurentide ice sheet on the east.
Arrows indicate directions of ice flow.

Legacy of the Ice Age

T HE WORLD was a very different place 20,000 years ago. Huge ice sheets, similar to the present-day Greenland and Antarctic ice sheets, covered northern Europe and the northern half of North America. The Laurentide ice sheet, comprising some 18 to 35 million cubic kilometres of ice, covered much of Canada, burying the land surface to depths of up to 3 km. To the west was the much smaller Cordilleran ice sheet, which covered nearly all of B.C. and southern Yukon Territory.

The Ice Age, or **Pleistocene** Epoch as geologists term it, lasted from about 2 million years ago until 10,000 years ago. Climate during the Pleistocene Epoch was generally cooler than during both the preceding Tertiary Period and the following **Holocene** Epoch. The Pleistocene was also a time of major climatic shifts on time scales ranging from decades to hundreds of thousands of years, resulting in the repeated growth and decay of continental ice sheets. Studies of cores of deep-sea sediments of Pleistocene age provide evidence for eight major climatic cycles in the last 800,000 years, each lasting about 100,000 years and each marked by sharp fluctuations in climate at shorter time scales. Climatic cycles of shorter duration (tens of thousands of years) occurred from before the beginning of the Pleistocene to 800,000 years ago. The colder parts of most of these climatic cycles were times of widespread glaciation in Canada.

Glaciation profoundly affected Canada. It changed the landscape of the country, as well as its vegetation and animal life. Sea level fluctuated up to 200 m above and below its present position owing to loading of the crust by glaciers and transfers of enormous quantities of water between oceans and ice sheets. The Laurentide and Cordilleran ice sheets, by virtue of their great thickness and extent, also radically altered oceanic and atmospheric circulation.

Mount Waddington (elevation 4019 m), the highest peak in the Coast Mountains. Glaciers flowing from Mount Waddington formed part of the vast ice sheet that covered British Columbia during the Ice Age.

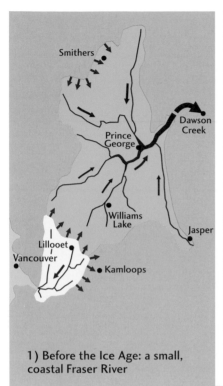

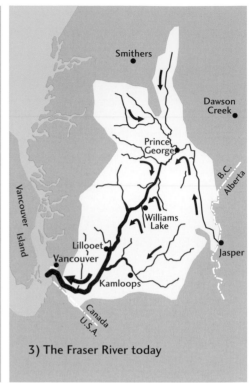

1) Before the Ice Age: a small, coastal Fraser River

2) During the Ice Age: expansion by stream piracy

3) The Fraser River today

Shifting drainages — piracy of the Fraser River

The Fraser River between Hope and just north of Lillooet is confined within a narrow, steep rock canyon. This canyon follows a major fault, the Fraser Fault, which is a major break in the Earth's crust. Crushed rock along the fault erodes easily, allowing rapid downcutting by the Fraser River. At a few places south of Boston Bar, the Fraser River is deeper than it is wide — very unusual for a river of this size. Normally, a river as large as the Fraser flows in a bigger, broader valley, has a lower gradient, and is much wider than it is deep.

Why does so large a river run through such a small valley? The answer may be that the Fraser River, at least as we know it, has not flowed in its present course for

*Three maps showing how the Fraser River **drainage basin** evolved over the last several million years. Prior to the Ice Age (panel 1), much of the present Fraser River basin was drained by a north- and east-flowing river system that may have flowed into the present Peace River. The Fraser River itself was small and had its headwaters north of the present town of Lillooet. During the Ice Age (panel 2), the Fraser River drainage basin grew at the expense of the north- and east-flowing river system. Stream erosion, volcanic activity, mountain building, and the repeated growth and decay of glaciers rearranged the drainage into its modern form (panel 3). The red arrows indicate encroachment of coastal rivers into the old north- and east-flowing river basin.*

very long. Perhaps as recently as a few million years ago, central B.C. was drained by an east- and north-flowing river system, quite different from the present south-flowing Fraser drainage. This river system drained through the Rocky Mountains, perhaps along the present course of the Peace River or perhaps along a route farther south. Evidence of this ancient river system can be seen in anomalies in the present pattern of drainage. Many tributaries to the Fraser flow northward, rather than southward as one might expect if the Fraser had always drained to the south through central B.C. Examples are the Quesnel and Blackwater (West Road) Rivers. In addition, ancient gravel and sand exposed in bluffs along the Fraser River between Prince George and Quesnel show evidence of having been deposited by a river that flowed north, opposite the present direction of flow.

Fraser River canyon north of Lillooet. The Fraser River flows in an unusually deep, narrow valley in this area.

It is likely that a separate smaller river flowed south through the Fraser Canyon into Puget Sound or the Strait of Georgia at this time. The **drainage divide** separating this smaller river from the headwaters of the larger north-flowing river system may have been located in the southern Chilcotin country near Gang Ranch.

In the last several million years, the small, south-flowing river, with its relatively steep gradient and therefore higher erosive capability, captured, or pirated, the headwaters of the northern drainage. It is not clear whether this capture occurred gradually over a period of hundreds of thousands of years or more rapidly. Nor is the cause of the capture known, although disruption of the drainage pattern by advancing or retreating Ice Age glaciers or the rise of the Rocky and Coast Mountains may have been important factors. It is likely, however, that the present Fraser River drainage was established in the last few million years, relatively recently in the geological scheme of things.

Cordilleran ice sheet

Glaciers in the high mountains of western Canada advanced as the Earth chilled early in each glacial cycle. Over a period of thousands of years, these glaciers advanced beyond mountain fronts and coalesced over lowlands and plateaus to create a vast sea of ice known as the Cordilleran ice sheet.

The last Cordilleran ice sheet, when fully developed, extended from central Yukon in the north to northern Washington, Idaho, and Montana in the south, and from the Pacific Ocean in the west to the Rocky Mountain foothills in the east. The surface elevation of the ice sheet was more than 2300 m over much of B.C., and ice was more than 2 km thick over major valleys. Glaciers flowing down **fjords** and valleys in the Coast Mountains covered large areas of the **continental shelf** west of the B.C. mainland. Glaciers originating in the southern Coast Mountains and in the mountains of Vancouver Island coalesced over the Strait of Georgia to produce a great lobe of ice that flowed to the south end of the Puget Lowland in Washington, terminating near Olympia. Ice streaming down valleys in south-central and southeastern B.C. likewise terminated as large lobes in eastern Washington, Idaho, and Montana.

The Homathko Icefield in the southern Coast Mountains about 150 km northwest of Whistler. The Lower Mainland looked something like this 16,000 years ago.

The ice-sheet phase of the last glaciation, referred to in B.C. as the **Fraser Glaciation**, occurred between about 20,000 and 12,000 years ago and was preceded by a lengthy period of glacier growth, during which glaciers at times extended into, but did not completely cover, the Strait of Georgia and Fraser Lowland. The ice sheet achieved its maximum extent on the south coast about 16,000 years ago and had completely disappeared 5000 years later.

Each glacial cycle ended with rapid climatic warming. At the end of the Fraser Glaciation, the Cordilleran ice sheet retreated from peripheral glaciated areas and thinned throughout much of the interior. Along the B.C. coast, glaciers retreated inland as sea level rose because of melting ice. In hilly and mountainous areas, the pattern of deglaciation was more complex, with uplands becoming free of ice first, dividing the ice sheet into a number of valley tongues that retreated separately in response to local conditions.

Ice cover in the Lower Mainland 16,000 years ago. See page 51 for an idea of what the area looked like 2500 years later.

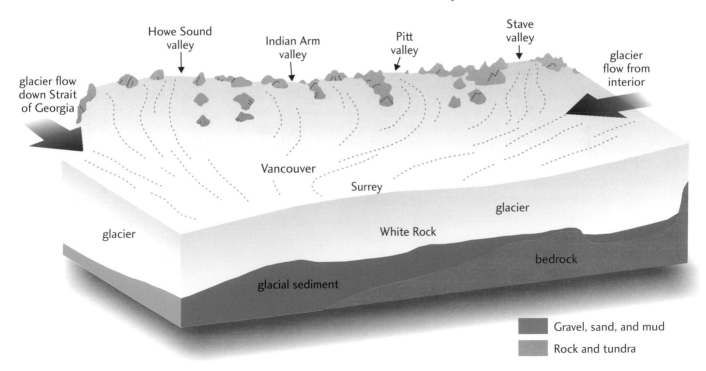

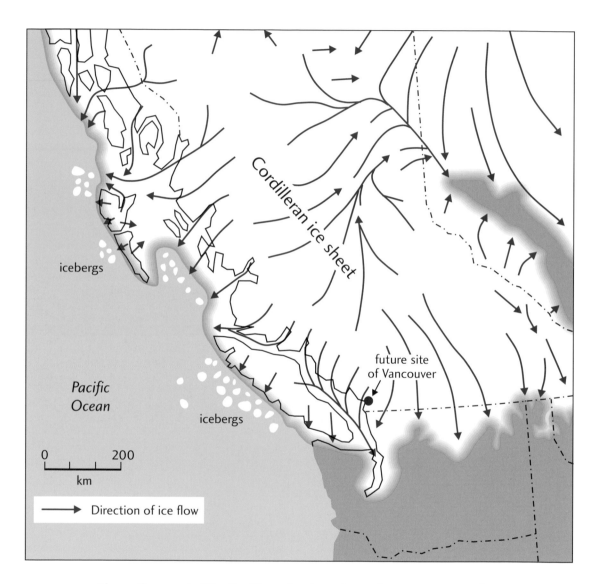

Cordilleran ice sheet

icebergs

Pacific
Ocean

icebergs

future site
of Vancouver

0   200
km

→ Direction of ice flow

The southern part of the Cordilleran ice sheet about 16,000 years ago at the
maximum of the last glaciation. The upper surface of the ice sheet reached up
to 3000 m above sea level. Arrows indicate directions of ice flow.

Glacial sculpting of the landscape

Evidence of Ice Age glaciation can be seen everywhere in B.C.'s landscape. Classic montane glacial erosional landforms, including **cirques** and sharp, narrow ridges (**arêtes**), abound in the higher parts of the southern Coast and Cascade Mountains, where some summits stood above the surface of the ice sheet as **nunataks**. In many places, the upper limit of the former ice sheet can be inferred from a change in the gross character of the landscape — ridges and peaks below this limit typically are rounded, whereas those above the limit are sharper and more rugged. The upper elevation of glaciation in the southern Coast Mountains is about 2000 to 2500 m; closer to the southern limit of the ice sheet in the Cascade Mountains at the B.C.-Washington boundary, it is lower, about 1800 m.

View of the southern Coast Mountains, showing typical glacially eroded landforms. Note U-shaped valleys, cirques, and arêtes.

The upper limit of the former Cordilleran ice sheet in the Cascade Mountains near the Canada-U.S. border southwest of Hope is etched in the mountains. Peaks that were above the limit of the ice sheet are sharp and rugged, whereas the lower ridges have been rounded by glacial erosion.

Lynn Canyon

Lynn Canyon Park offers tantalizing clues about a time when stream courses on the North Shore were different from those of today. If heights don't bother you, walk out on the suspension bridge over the narrow bedrock canyon of Lynn Creek, its waters rushing down to Burrard Inlet. To geologists, Lynn Canyon is very unusual. Over time a stream should have widened this canyon, creating a valley more typical of those occupied by streams and rivers elsewhere in the region. Geologists thus suspect that Lynn Creek has not flowed in Lynn Canyon for very long (in the geological sense of the word "long").

Just upstream of the suspension bridge, Lynn Creek flows in a more typical, broad valley. Take the trail on the western side of Lynn Creek and compare this broad valley reach, which has no bedrock exposure, with the strikingly different bedrock canyon to the south. About 200 m upstream (north) of the suspension bridge, you reach a point where Lynn Creek abruptly leaves another bedrock canyon, similar to that at the bridge, and enters its broad reach. "Very strange," says the geologist, "How can this be explained?"

The most likely explanation is as follows. The broad reach represents a cross-section through a partly sediment-filled, former valley of Seymour River. The buried valley extends in a northeast-southwest direction from Seymour Valley to Burrard Inlet to the west. Lynn Creek cuts across it at an angle. Since the present course of Seymour River became established to the east, the old valley was abandoned and became filled with sediment. You can see remnants of this sediment fill as you walk along the valley bottom between the canyon sections.

"All this is fine," you say, but "how did Lynn Creek cut canyons in rock?" "Very good question," replies the geologist. Near the end of the Ice Age, some 13,000 years ago, Lynn Canyon was a very different place. Glaciers had just retreated from the area, and streams flowing from the North Shore mountains entered the sea at what is now North Vancouver at a time when sea level was 200 m higher than today (page 48). Large deltas were built out into the high sea at the mouths of meltwater streams on the North Shore, burying the pre-existing landscape. Soon, however, the sea began to fall, and streams, including Lynn Creek, started to cut down through their deltas. Lynn Creek soon encountered high rocky ground flanking the buried valley at the present site of Lynn Canyon Park — in other words, the creek was "let down" onto rock. The creek sought the lowest points it could find on the rock surface. Once it found these points, its course was fixed and all it could do was cut down through the rock. Over the next 13,000 years it carved the present canyon.

Another question you might ask is "How old is the old, buried valley?" We are not able to answer this question exactly, but we do know that the valley is at least 60,000 years old and probably much older. A thick bed of **peat**, carbon-dated at 30,000 to 40,000 years old, is present on the floor of the old valley just north of suspension bridge. The peat rests on glacial till that is more than 60,000 years old. Because the till lies on the floor of the valley, we can logically conclude that it is younger than the valley itself. You can see both the peat and the till on the western side of the valley next to the elevated walkway at the base of the stairs leading down from the suspension bridge.

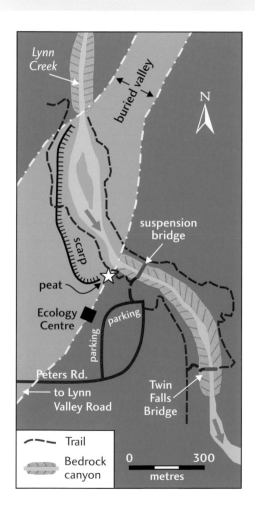

43

Classic, glacier-carved, U-shaped hanging valley, Milford Sound, New Zealand. Water plunges from the floor of the hanging valley to the main valley, now an ocean fjord.

Most mountain valleys in B.C. have U-shaped cross-sections characteristic of valleys that were eroded by glaciers. Exceptions include canyons cut by rivers since the Ice Age, for example Lynn and Capilano Canyons on the North Shore.

In contrast, mountain valleys in areas that have never been glaciated typically have V-shaped profiles produced by river erosion. Some of the smaller valleys in the Coast Mountains "hang" above U-shaped trunk valleys to which they are tributaries. An example is the valley of Shannon Creek near Squamish. Because the much larger trunk glacier flowing south down Howe Sound eroded more rapidly and deeply than the smaller tributary glacier, a **hanging valley** was left high on the wall of the fjord. Shannon Creek now tumbles 335 m over the boundary where the smaller glacier used to join the larger one.

The coastline of B.C. is indented by fjords, which are old river

valleys that have been deeply
eroded by glaciers and are now
flooded by the sea. Fjords mark
the paths of fast-flowing, highly
erosive, outlet glaciers of the
former Cordilleran ice sheet. B.C.
fjords extend up to 150 km inland
and have water depths of as much
as 750 m. Howe Sound and Indian
Arm are two examples close to
home. The fjord coastline of B.C.
is among the grandest in the
world, surpassing the coastlines of
Norway and New Zealand and
rivalling that of Greenland.

Howe Sound, a typical B.C. fjord.
This steep-walled inlet is a Tertiary
river valley that was deepened and
widened by glaciers during the Ice Age.
The town of Squamish is located on
the Squamish River delta at the head
of the fjord.

Although they do not qualify as
fjords, the Strait of Georgia and Juan de Fuca Strait have also been
eroded by glaciers. Were it not for glaciation, these water bodies
would be much shallower than they are today. Glaciers also deepened valleys on the southern flank of the Coast Mountains north
of the Fraser Lowland. When the glaciers disappeared, **fjord lakes**
such as Pitt, Alouette, and Harrison occupied the valleys.

Lowlands bordering the Strait of Georgia have been eroded
and streamlined by glacier ice. Much of Vancouver's west side, for
example, is located on a gently ridged surface with a distinct
southeast grain, the result of erosion by southeast-flowing ice.

Not all features produced by glacial erosion are large. Some
bedrock has been polished and striated by glaciers. Glacier ice is
too soft to scour solid rock on its own, but stones and sand grains
embedded in the ice act like sandpaper as they are dragged across
the rock surface. These objects both scratch and smooth the rock.
Glaciers can also "pluck" large blocks from highly fractured rock,
especially where such rock rises as ridges above the surrounding
terrain. The blocks are frozen into the bed of the glacier and slowly
carried away. Repeated action of this sort may give rise to large
rock ridges that are smooth at their up-ice ends but have ragged
down-ice terminations. Such rock ridges are common in the

Glacially polished and striated granite at the base of Stawamus Chief near Squamish.

Cheakamus River valley north of Squamish; you can see them when you drive along Highway 99 in the vicinity of Brohm Lake.

Glacial plucking also has probably played an important role in creating the steep western face of Stawamus Chief near Squamish. The "Chief" (page 24) reminds one of Half Dome in Yosemite, and in fact their geology is similar. Both consist of granitic rocks and both have been shaped by glacial erosion. The rocks of the Chief are fractured along steep **joints** that run parallel to the rock face. Glaciers flowing down Howe Sound during the Ice Age plucked slabs of granite along these joints and created the steep rock face you see today.

Stop at the Stawamus Chief parking lot, located 2.5 km south of Squamish on Highway 99, and have a look at this awe-inspiring rock face. You will likely see climbers during good weather; from the parking lot they look like ants. On the right side of the face a black basaltic dyke cuts vertically up the wall. At the south end of the parking lot is a beautiful example of glacially polished and scratched granite.

Ice Age sediments

Vast amounts of rock and sediment were removed by glaciers and streams from the Coast and Cascade Mountains during the Ice Age. What became of all this material? Much of it found its way to the Pacific Ocean, but large amounts of sediment were deposited on coastal lowlands and in the Strait of Georgia. The Fraser Lowland, for example, is underlain by a blanket of Ice Age sediments up to several hundreds of metres thick. These sediments mask the buried bedrock surface, although in some places bedrock hills poke through the sediment cover (for example, Burnaby and Sumas Mountains).

Ice Age sediments in the Fraser Lowland comprise a variety of materials of different origins. Most of the sediments are layered gravel, sand, and silt deposited by melt-water streams near the margins of glaciers that advanced into the Fraser Lowland from adjacent mountains. The gravel and sand are important resources, used in construction and landscaping (see page 163). Till, deposited directly from glacier ice, is also present but is a relatively small part of the Ice Age sediment package. Good examples of till can be seen in many building excavations in Vancouver, along the Cypress Bowl road, and at the top of the sea cliffs at the University of British Columbia.

The youngest Ice Age sediments at Vancouver are products of the most recent, or Fraser, glaciation. Thick bodies of sand deposited by streams flowing off advancing glaciers during the early stage of the Fraser Glaciation are termed Quadra Sand by geologists. Quadra Sand was overridden and eroded by the Cordilleran ice sheet at the peak of the Fraser Glaciation. Before retreating, however, the ice sheet left a layer of till on top of the sand. Quadra Sand is exposed in many sea cliffs around the Strait of Georgia, for example at Point Grey in Vancouver, and is also an important **aquifer.**

Till, a glacial deposit consisting of stones in a matrix of clay, silt, and sand, along the road to Cypress Bowl ski area near Highview Lookout. The largest stones are about 15 cm long.

Point Grey — cliffs of Ice Age sand

The beach at Point Grey is one of our favourite spots — an oasis of calm and beauty in the busy city. But it is worth noting that the beach is "clothing optional," something you may wish to take into account if you visit "Towers Beach" on a warm, sunny day.

The sea cliff above the beach provides a glimpse back into the Ice Age. The cliff is composed mainly of sand, with some silt and peat beds below 18 m elevation. The sand was deposited by streams that flowed from glaciers advancing south along what is now the Strait of Georgia and Howe Sound about 27,000 years ago at the beginning of the Fraser Glaciation. At that time, much of the Strait of Georgia and Howe Sound north and northwest of Point Grey were covered by glaciers, much like parts of southeastern Alaska today. Eventually, the glaciers overrode Vancouver and Point Grey, entombing the entire Lower Mainland under about 2 km of ice. A thin layer of bouldery till at the top of the sea cliff records the passage of the glaciers over the area.

The Point Grey sea cliff is vulnerable to attack by waves and currents, and much money has been spent protecting it from erosion. In 1980 and 1981, a 30- to 40-metre-wide berm of cobble gravel was placed on the beach between the two World War II concrete observation towers. Ribs of coarse crushed rock were laid down perpendicular to the shore to anchor the berm and protect it from erosion. A fence was erected at the top of the cliff to discourage people from climbing the bare slopes. An attempt was made to revegetate the scarred sections of the cliff by spraying a mixture of grass seed and fertilizer from aircraft. Trees at the top of the cliff were cut down to prevent the protective soil mantle from being stripped as a result of toppling.

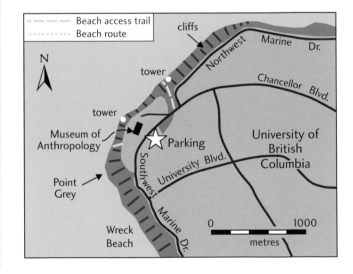

till

Sea cliff at Point Grey, Vancouver. The cliff exposes sand and silt (the latter are dark layers near the base) deposited by meltwater streams that flowed from an advancing glacier about 27,000 years ago during the Ice Age. The sand is capped by a thin layer of till. The photo was taken in 1972 before measures were taken to protect the cliff from erosion. The steep slope and patchy vegetation are evidence of rapid erosion at that time. Since then, vegetation has become established on much of the cliff.

Have these works been successful in reducing cliff erosion? The answer is a qualified yes. The cliff, which was bare and retreating rapidly in the 1970s, is now vegetated and more stable. The slopes, however, are still steep and, over time, the cliff top may retreat several tens of metres, endangering the Museum of Anthropology. In addition, the protective berm has been severely eroded in the twenty-two years since it was placed on the beach. Much of the gravel has been carried away by waves during storms, exposing the ribs of coarse crushed rock that extend out from the shore like orphans. In a few places, waves have nearly breached the berm and are poised to attack the base of the cliffs once again.

Is there another solution to the erosion problem at Point Grey? Allowing the cliff to retreat naturally is unacceptable because university buildings, including the Museum of Anthropology, would be lost. A possible "soft engineering" solution is to supply the beach at Point Grey with sand dredged from the North Arm of the Fraser River. The sand would increase the width of the beach, causing waves to break farther from the base of the cliffs. Of course, the sand would be eroded, necessitating continuing replenishment, but a sandy beach would have more recreational value than a cobbly one, and anyway sand is currently dredged from the North Arm and dumped in the Strait of Georgia. Perhaps it could be put to better use!

Porteau Cove — a giant undersea Ice Age ridge

A half-hour drive north of Horseshoe Bay on Highway 99 brings you to Porteau Cove Provincial Park. You can't see it, but just offshore of Porteau Cove is a huge, broad ridge of gravel and till that loops across Howe Sound. The ridge reaches to within 30 m of the sea surface, whereas the seafloor to the south and north is more than 200 m deep. This ridge is a **moraine** built 13,000 years ago by a glacier flowing south down Squamish valley past what is now Squamish to terminate in an iceberg-infested sea at Porteau Cove. The cove is a popular diving spot and a marine park. Were it not for the shallow moraine, water depths would be too great for diving.

Erratic in Coquitlam left by melting ice about 14,000 years ago.

Thick, bedded silt of the same age as Quadra Sand is present in the lower reaches of some mountain valleys bordering the Fraser Lowland, notably the Capilano, Coquitlam, and Chilliwack valleys. The silt was deposited in large lakes impounded at the margins of glaciers flowing down the Strait of Georgia and across the Fraser Lowland during the Fraser Glaciation. The glaciers blocked the mouths of the mountain valleys and thus dammed the streams that flowed out of them.

Have you ever noticed the huge boulders, mainly granite, scattered around Pacific Spirit Park, Coquitlam, White Rock, and other upland areas of the Lower Mainland? Where did these boulders come from and how did they end up where they are? They were not trucked in; imagine the size of truck needed to move some of these immense boulders. Rather, the boulders were carried on or within glaciers that flowed from the Coast Mountains across the Fraser Lowland during the Ice Age. As the glaciers melted, the boulders were released from the ice and deposited where we see them

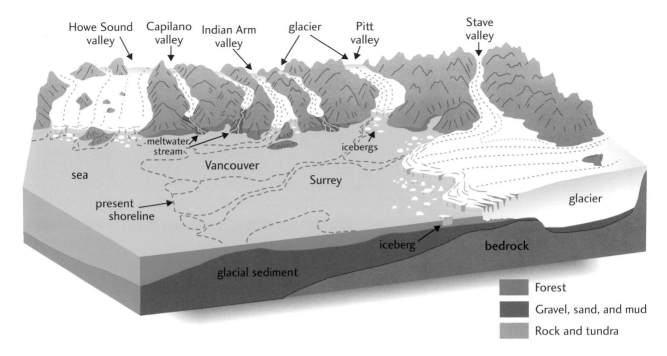

The following labels appear on the diagram: Howe Sound valley, Capilano valley, Indian Arm valley, glacier, Pitt valley, Stave valley, meltwater stream, icebergs, sea, Vancouver, Surrey, present shoreline, iceberg, bedrock, glacial sediment, glacier

Legend:
- Forest
- Gravel, sand, and mud
- Rock and tundra

today. Geologists call these boulders **erratics**. Some of the erratics were deposited directly on dry land from the melting ice. Others, however, may have dropped to the seafloor from melting icebergs.

About 15,000 years ago, glaciers began to retreat northward up the Strait of Georgia and eastward across the Fraser Lowland. The land was still depressed by the weight of the remaining ice, and the sea flooded newly deglaciated terrain below about 200 m elevation around Vancouver. Most of what is now the western Fraser Lowland was a bay of the Strait of Georgia 13,500 years ago; only the top of Burnaby Mountain rose above the sea as an island. The bay was surrounded by tidewater glaciers and was filled with icebergs, much like parts of Glacier and Yakutat Bays in Alaska today. Meltwater streams flowed across barren gravel plains into the bay. Marine and deltaic sediments accumulated on the submerged lowlands at this time. These sediments include the large sand and gravel deltas on

The Lower Mainland during retreat of the Cordilleran ice sheet about 13,500 years ago. As glaciers retreated, the sea flooded land depressed by the weight of the ice. A bay formed in the western Fraser Lowland, surrounded by tidewater glaciers and filled with icebergs. Muddy sediments were deposited on the seafloor up to the present elevation of 200 m. Large deltas of sand and gravel formed at the mouths of valleys where meltwater streams met the sea. See page 39 for an idea of what the area looked like 2500 years earlier.

In this photo, taken in 1961, Dawes Glacier flows into the head of Endicott Arm in southeastern Alaska and calves off numerous icebergs into the sea. The walls of the inlet have been steepened and polished by the erosive action of the glacier. The high peaks in the background are near the boundary between B.C. and Alaska. Howe Sound looked like this about 14,000 years ago.

which part of North Vancouver has been built. Thick **glacial-marine** silt and clay, locally containing clam and snail shells and stones dropped from melting icebergs, underlie much of the western and central Fraser Lowland, including Langley, Haney, and Aldergrove, and also are present beneath the seafloor in the Strait of Georgia and in fjords such as Howe Sound. These sediments are thickest in the central Fraser Lowland where they were deposited on the seafloor near the margin of a tidewater glacier. Glacial-marine sediments in the Strait of Georgia and Howe Sound lie beneath up to many tens of metres of silt and clay that have accumulated in these basins over the last 12,000 years.

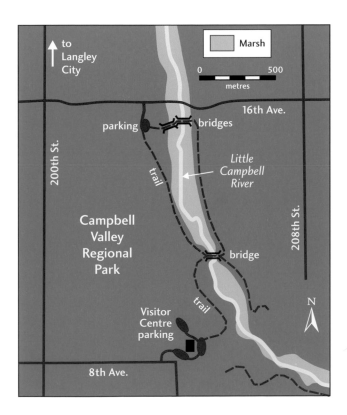

Campbell Valley Regional Park

Campbell Valley Regional Park , south of Langley, is a wonderful area for a family outing; be sure to take a picnic lunch. But the valley may puzzle you. If Little Campbell River seems much too small to have carved the large, marsh-filled valley in which it flows, you are correct. The valley is a relict of the last days of the Ice Age about 13,000 years ago, when it carried meltwater from the glacier in the Fraser Lowland east of Aldergrove west to the sea. Just northwest of the park, the meltwater stream met the sea and built a large delta. At the time, sea level was 30 m higher than it is today. The flat gravelly surface on which the community of Brookswood (part of south Langley) is built is the surface of this ancient delta.

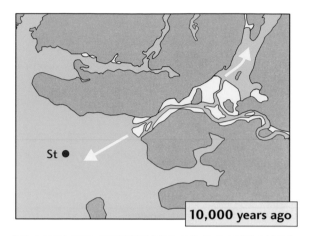

10,000 years ago

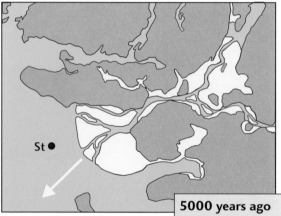

5000 years ago

Growth of the Fraser delta and floodplain, shown in yellow, over the last 10,000 years ago. Yellow arrows show the main direction of sediment accumulation and growth of the Fraser delta. Richmond and Delta are located on land that didn't exist 10,000 years ago. What might this area look like 5000 years from now?

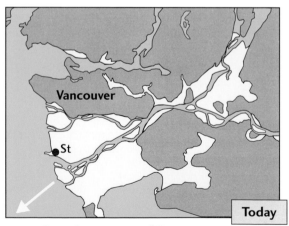

Vancouver

St

Today

St ● Present site of Steveston

After the Ice Age

THE LAST of the Ice Age glaciers retreated out of the Fraser Lowland back into mountain strongholds about 12,000 years ago. The climate has changed since then, and the prolonged severe chill of the Ice Age has not returned. Much of the landscape 12,000 years ago was little different from that of today, but the floodplain and delta of the Fraser River did not exist. An arm of the sea extended eastward from the Strait of Georgia along the present course of the Fraser River to the Abbotsford area and also reached into Pitt Lake. Plants and animals rapidly colonized the sterile, newly deglaciated landscape, but for many thousands of years the flora and fauna were fairly different from those at present. Humans may have reached the south coast as early as 13,000 years ago and have continuously occupied the region ever since. In contrast, the first settlers of European stock arrived less than 200 years ago.

Climate and vegetation change since the Ice Age

The climate on the south coast during deglaciation was cool and moist. Freshly deglaciated ground was colonized by shrubs, herbs, mosses, and lodgepole pine. Subalpine fir and spruce accompanied or closely followed pine. As the climate warmed, vegetation changed rapidly and new species migrated in from other areas. The cool, moist, deglacial climate was replaced by a warm, drier one. By 10,000 years ago, climate was probably about 1° C warmer than today. Lodgepole pine forests were replaced at lower elevations by forests dominated by Douglas fir, western hemlock, and, locally, alder. Spruce, fir, and other trees continued to be important elements of forests at higher elevations. Forest fires were probably very common during this warm, dry period.

Some time between about 9000 and 7000 years ago the climate began to cool. Cooling probably was not sudden, but rather took place over hundreds or even thousands of years, with much variability over periods of decades and centuries. By 4000 to 5000 years ago, modern coastal forests, characterized by western red cedar, hemlock, and Douglas fir were established under a climate not too different from that of today (page 14).

Glaciers in western Canada and other mountain regions have advanced on several occasions during the last 5000 years. None of these advances compares to the massive expansion of glaciers during the Ice Age, but cooling between the thirteenth and nineteenth centuries was significant enough for scientists to have termed the period the **Little Ice Age**. Many people are surprised to learn that the maximum advance of glaciers during the last 10,000 years occurred only 150 to 250 years ago near the end of the Little Ice Age.

Changing climate in southwestern B.C. from the end of the Ice Age to today. Climate change is depicted on two time scales — the last 10,000 years (bottom) and the last 1000 years (top). Most scientists think that the sharp warming trend of the last 20 years is due, at least in part, to rising concentrations of greenhouse gases in the atmosphere caused by fossil fuel burning

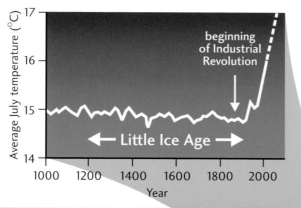

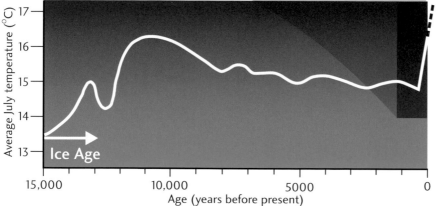

The rise and fall of the sea

As mentioned above, the sea surface at Vancouver at the end of the Ice Age was up to 200 m higher than it is today. This is a consequence of loading, or weighting, of the Earth's crust by the Cordilleran ice sheet. With the disappearance of the ice sheet, the land rapidly rose. Water was entering the oceans from melting ice sheets, but global sea-level rise was outstripped on the southern B.C. coast by rebound of the land. As a consequence, the shoreline at Vancouver fell rapidly as deglaciation progressed. The rate of sea-level fall is impressive: almost 1 metre every 10 years, or 10 cm per year from 13,000 to 12,000 years ago. At such a pace, anyone living at Vancouver at that time would have been able to watch the sea retreat.

Glaciers' last stand?

Did you know that most glaciers in the Coast Mountains were larger in the nineteenth century than at any other time in the last 10,000 years? The evidence is striking and is familiar to anyone who spends lots of time near glaciers. The next time you are hiking in the mountains, note the unvegetated or poorly vegetated zones just outside present-day glacier margins. The sharp break between the unvegetated zone and the mature forest outside or above it is called a **trimline**. Trimlines mark where glaciers stood in the mid-nineteenth century at the end of the Little Ice Age. In the twentieth century, glaciers have retreated in response to a warming climate. Warming has increased since about 1980, perhaps because of human-induced increases in carbon dioxide and other greenhouse gases in the atmosphere.

"Icemaker Glacier" in the southern Coast Mountains northwest of Pemberton, showing evidence of recent retreat. The trimline separating vegetated slopes above from unvegetated slopes in the foreground marks the position of the glacier in the nineteenth century.

Why are there sea shells in my back yard?

Many Lower Mainland residents who live far from the seashore have found sea shells while digging in their yards. They are naturally puzzled and wonder how shells can occur so far from the shore and so high above present sea level. Perhaps, they muse, seagulls dropped them or maybe people left them there. Neither of these explanations is correct. The shells include clams, snails, and other marine organisms that are some 12,000 to 14,000 years old and lived on or just below the seafloor as Ice Age glaciers retreated from the area. When these animals died, their shells were preserved in mud and sand that accumulated on the seafloor. The marine sediments and their entombed fossils were later elevated above sea level as the land rose due to removal of the ice that had been loading the crust, and they were left high and dry, far from the present shore.

Fossil marine clams and snails, about 14,000 years old, found in marine sediments at an elevation of 65 m along the Vancouver Island Highway north of Parksville. Top row, left to right: butter clam (Saxidomus giganteus), Greenland cockle (Serripes groenlandicus). Centre, smooth pink scallop (Chlamys rubida). Bottom row, left to right: Arctic moon snail (Natica clausa), Lyre whelk (Neptunia lyrata). For scale, the scallop in the centre is 6 cm across.

Evidence for low sea levels during the Holocene

The sea continued to fall relative to the land for several thousand years after deglaciation. By 9000 years ago, sea level at Vancouver may have been 10 to 20 m below the present shore; remarkably, areas that are now shallow seafloor were probably dry land at that time. Scientists have found fossil land plants, carbon-dated at about 9000 years old, in sediments up to 15 m below sea level at Pitt Meadows, Sumas, and Port Mann in the Fraser Lowland. Additional evidence for lower sea levels includes shoreline features such as beaches, **spits**, and deltas at water depths of 50 m or more in Juan de Fuca Strait.

The sea began to rise from its low stand about 8000 to 9000 years ago, due to melting of the remnants of the Laurentide ice sheet and thermal expansion of oceans. The local rate of rise in sea level was rapid at first but gradually decreased around 7000 years ago. By 6000 years ago, sea level was only about 5 m below its present position at Vancouver, and over the last 3000 years, it has

View north across Pitt Meadows towards Pitt Lake and the southern Coast Mountains. At the end of the Ice Age, Pitt Lake and the lowland in the foreground were an arm of the sea extending eastward from the Strait of Georgia. The floodplain in the foreground formed over the last 10,000 years as the Fraser River built a delta northward into Pitt Lake, isolating it from the sea.

Pitt fjord

At the end of the Ice Age, Pitt Lake was a fjord that extended east and north from the Strait of Georgia past what is now New Westminster. The Fraser River advanced westward towards New Westminster, filling in the mouth of the fjord located in the Pitt Meadows area, thus creating Pitt Lake (see page 54).

Fort Langley

Fort Langley was B.C.'s first capital, established in 1827, long before the cities of Victoria and Vancouver existed. Most people visit the fort, which is a National Historic Site, unaware of the site's geological significance. The old fort was built on a small patch of ground elevated above the low area on which the present community of Fort Langley is located. This site was chosen because the drainage is good, flooding by Fraser River is not a problem, and perhaps because the settlers could see some distance up and down the river from the slightly elevated position of the fort. The bench on which the fort is located is underlain by sandy deltaic sediments that were deposited by the Fraser River about 11,000 to 12,000 years ago. At that time, the river flowed into an arm of the sea that extended east from the present Strait of Georgia past New Westminster into the Pitt Meadows and Maple Ridge area. Since then, the Fraser River built its delta and floodplain westward and the sea fell to and then below its present level. The old, elevated delta plain was incised and eroded by the Fraser River; all that is left today is the remnant surface on which the fort was built and a few other benches farther upriver.

This remnant surface is striking from the air. It lies between the present-day Fraser River and an older cutoff channel of the river that loops around Fort Langley on its south side. No one knows when the river abandoned the old channel and began to flow along its present course, but it was probably thousands of years ago.

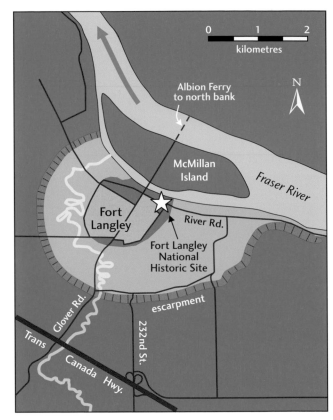

Former Fraser River channel

The Fraser River once flowed in a loop south of Fort Langley, but this channel was abandoned when the river shifted to the north. Since then, the former river channel has filled with silt and sand and become a flat floodplain.

not varied more than a metre or two from its current level.

The wild gyrations in sea level outlined above have affected the evolution of the lower Fraser River and its floodplain. During deglaciation, the Fraser River did not exist because its path was blocked by glaciers in the eastern Fraser Lowland, the Fraser Canyon, and in the southern interior of B.C. A river flowed off the decaying glacier margin near Abbotsford into an arm of the sea extending eastward from the Strait of Georgia.

When ice disappeared from the Fraser Canyon and the B.C. interior, the Fraser River became established in its present course. It rapidly built its delta and floodplain westward into the sea, past what is now Haney, Fort Langley, and Pitt Meadows (page 54). The seaway that extended up the Fraser Valley to Abbotsford was rapidly filled with silt, sand, and gravel eroded from glacial sediments. The glacial sediments were not well vegetated and were easily eroded and washed into streams of the newly formed Fraser River drainage. The early westward advance of the Fraser River occurred while the sea was falling relative to the land. Later, less than 8000 years ago, the river was forced to **aggrade**, or build up, its floodplain to keep pace with the rising sea. The end result is that the Fraser River has completely filled the seaway that formerly lay in its path.

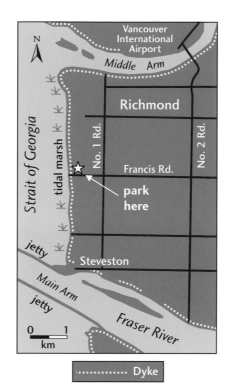

Lulu Island dyke

Drive to the sea dyke at the end of Francis Road in Richmond and walk south along the **dyke** to Steveston. As you stroll along, the Lulu Island tidal marsh lies to your right (west) and the dyked Fraser River delta plain to your left (east). Seaward of the marsh are the flat, unvegetated flats of the Fraser delta, which extend another 6 km to the west before dropping off into deep waters of the Strait of Georgia. Lulu Island is located just north of the Main Channel of the Fraser River, which flows past Steveston to Sand Heads before mixing with the waters of the Strait of Georgia. The narrow fringes of marsh at the western and southern fronts of the Fraser delta are essential habitat for migrating birds and juvenile salmon. But this habitat is threatened by development and is at risk from rising sea levels, which may drown the marshes or squeeze them against the dykes. The sea dykes at the landward limit of the marshes protect Richmond and Ladner from flooding during high tides. The pump station at the end of Francis Road is one of several that remove water from the ditches that drain Richmond and pump it into the sea.

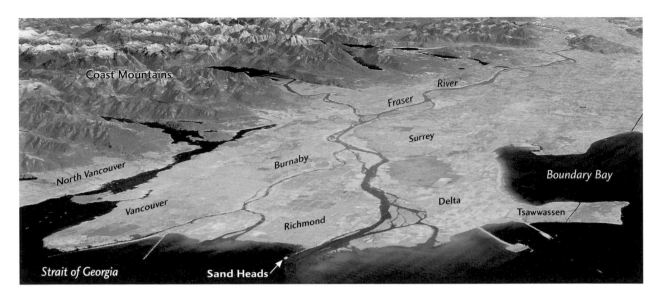

Computer-generated image of the Fraser delta and Fraser Lowland; view to the east.

Cut-away block diagram showing the geology of the Fraser delta and the natural processes operating at the delta front. The diagram also shows the important economic infrastructure situated at the delta front, including the Vancouver Airport, Deltaport, and the Tsawwassen Ferry Terminal. The block diagram overlaps a satellite image of the delta.

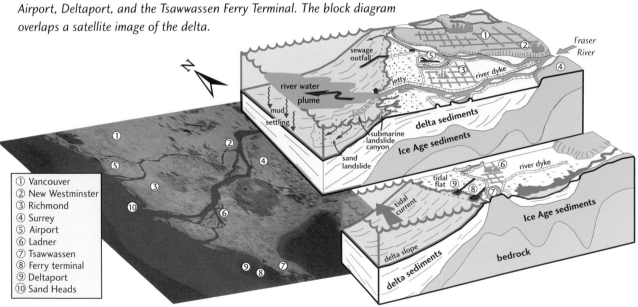

① Vancouver
② New Westminster
③ Richmond
④ Surrey
⑤ Airport
⑥ Ladner
⑦ Tsawwassen
⑧ Ferry terminal
⑨ Deltaport
⑩ Sand Heads

The Fraser delta — Mother Nature's dump

Over the last 10,000 years, a great pile of sediment has accumulated as a delta where the Fraser River flows into the Strait of Georgia. About 10,000 years ago, the Fraser River flowed into the sea at New Westminster. Since then, it has built its delta about 25 km seaward, creating the low-lying plain on which Richmond, Delta, and Ladner are located. Point Roberts peninsula, once an island in the Strait of Georgia, became connected to the mainland several thousand years ago.

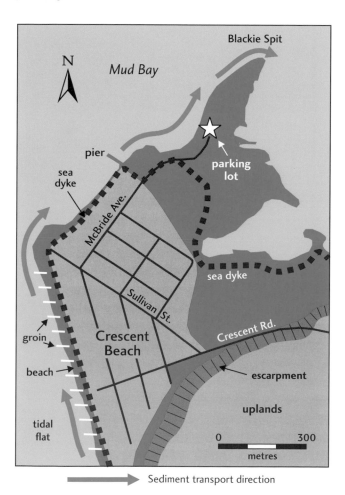

Sediment transport direction

Crescent Beach

This is a nice spot to go if you're interested in beaches and coastal processes. The community of Crescent Beach is located on a spit, an elongate, flat patch of sand and gravel that is very close to sea level. The spit has been built by the northward flow of ocean currents, which carry sand and gravel eroded from a coastal bluff just south of Crescent Beach. Most of the inhabited part of the spit is vegetated and stable, and the community is protected from storm waves at very high tides by sea dykes. The north end of the peninsula (Blackie Spit) is actively building into Mud Bay. As you walk along the beachfront at Crescent Beach, note the small jetties and groins built to trap sand and create a broader beach. Can you tell from the buildup of sediment on one side of these structures and erosion on the other side which way the currents are directed? What might happen to Crescent Beach if sea level were to rise one metre, as some scientists predict will happen over the next 100 years?

Greater Vancouver as seen from space, July 30, 2000. This image was taken from NASA's Landsat 7 satellite, which circles the Earth at an altitude of 705 km. The image shows the southern part of the Strait of Georgia, the Gulf and San Juan Islands, and the eastern end of Juan de Fuca Strait. Urbanized areas have a light grey colour.

The Fraser River discharges sediment-laden water (light blue) into the Strait of Georgia, where it is dispersed by currents and settles to the seafloor. This plume of muddy water is particularly pronounced during the spring and early summer, when Fraser River discharge is greatest. Silt and sand (pinkish grey) are deposited on the tidal flats at the river mouth; over time this sediment builds up and the delta front migrates westward.

Most of Greater Vancouver is built on Ice Age sediments and the Fraser delta. The delta soil is highly fertile and supports many farms (light green). Burns Bog is the conspicuous dark, egg-shaped patch of ground south of the Fraser River. In the top part of the photo, the Coast Mountains limit growth of the city to the north, and their steep mountain valleys provide much of Vancouver's water.

The light blue sediment plume extending down Howe Sound at the top of the photo comes from the glacier-fed Squamish River at the head of the inlet.

Reproduced with the permission of the Galiano Conservancy; the image is available as a poster from the Conservancy.

The Fraser River transports large amounts of sand along its bed. Some of the sand reaches the river mouth, where it accumulates and occasionally slumps into deep water in the Strait of Georgia. A submarine canyon off the main arm (Main Channel) of the river just west of Sand Heads is the pathway the sand follows to deep water. Most of the sand, however, is dredged to maintain a navigable shipping channel downstream from New Westminster.

Silt and clay suspended in the river waters accumulate in tidal marshes and on the seafloor off the delta front. Before the construction of river **dykes** in the 1900s, floods regularly deposited silt and clay on the flat Fraser delta plain. Now protected from floods by the dykes, the delta plain no longer receives new sediment from the river. At the same time, the thick pile of mud and sand underlying the delta plain is compacting, causing the land surface to slowly subside. Because much of the delta plain is already below the upper limit of tides, slow subsidence will pose an increasing problem in the future for people who live on it, for example in Richmond and Ladner. The problem will be magnified if a warming climate causes sea level to rise, as many scientists now think will happen. We may have to raise and strengthen the dykes that keep the sea out of Richmond and Ladner. Also, because the **water table** under the delta is very close to the surface, we may have to improve the pumps and drainage channels in Richmond and Ladner.

The Fraser River transports most of its sediment load during the late spring and summer when river discharge is high. Very little sediment reaches the river mouth during the winter. The high summer loads of the Fraser River are obvious to anyone flying over Vancouver during late May, June, or July (see photo on page 64). A brown plume of silt and clay can be seen extending west from the mouth of the Main Channel and North Arm far out into the Strait of Georgia and English Bay. At very high flows, the plume may reach the southern Gulf Islands and extend into Howe Sound. The plume is typically a 1 to 3 m thick layer of silt- and clay-laden, fresh to brackish water that overlies denser, more saline waters of the Strait of Georgia.

When rivers flood

FLOODING is a natural process that occurs along all rivers and streams. Rivers normally flow within their channels, but occasionally they overflow their banks and wreak havoc on the surrounding countryside. In the Fraser Lowland, low flatlands adjacent to the Fraser River and its tributary streams are vulnerable to flooding .

Floodplains owe their existence to repeated seasonal floods that spread silt-laden waters over large areas. Before European settlement, the mighty Fraser River was untamed — it regularly overtopped its banks and covered the swamps and grasslands of Hatzic Prairie, Pitt Polder, and the Fraser delta with its muddy waters. Beginning in 1864, dykes were built along the Fraser River and its tributaries to protect people and property from flooding. Since the last big Fraser River flood in 1948, the provincial and federal governments have spent $160 million building and improving over 500 km of dykes in the Fraser Valley. These improvements have allowed more than $13 billion worth of businesses and homes to be built on the floodplain behind the dykes. The present dyke system should withstand a flood as large as the "flood of record," which occurred in 1894. However, there is a one-in-three chance that a flood of this size or larger will occur in the next 60 years. Communities at risk from such a flood include Richmond, Delta, Langley, Coquitlam, Port Coquitlam, Pitt Meadows, Abbotsford, Chilliwack, Kent, and Mission. Nearly 350,000 people live on the lower Fraser River floodplain, an increase of 150,000 over the past 20 years. The risk of flood damage will only grow as the population increases.

Levels reached by Fraser River floods on a barn near Hatzic in the eastern Fraser Lowland.

Floods are repetitive phenomena. The risk that a flood of a particular size will occur along a river is determined from records of past discharge that commonly span many tens of years (see figure on this page). On the basis of these records, engineers and planners determine the annual probability, or statistical likelihood, that a certain-size flood will occur. For example, a flood of magnitude "x" may have an annual probability of 10 percent, meaning that, on average, it has a one in ten chance of occurring in any given year. In contrast, a flood twice as large as "x" may have an annual probability of only 1 percent.

In B.C., river dykes are typically designed to withstand a flood that has an annual probability of 0.5 percent; in other words, it is likely to recur, *on average,* once every 200 years (the so-called "200-year flood"). We emphasize "on average" — a 0.5 percent annual probability does not mean that the river will actually reach its 200-year level only once in 200 years. Rather, it is possible that the river will produce a flood of this size two times, three times, or not even once, in any 200-year period. It is *most likely,* however, that the river will flood to this level once every 200 years. The bottom line — don't assume that if a huge flood occurs this year it can't happen again next year.

Relation between the magnitude and probability of a flood on the lower Fraser River, based on data for the period 1912 to 2000. This plot shows that larger floods (shown as large circles) are much rarer than smaller ones (the smaller circles).

Floods — they're not all the same!

Floods in the Lower Mainland are caused by heavy rainfall and rapid snowmelt. Fraser River floods occur only during late spring and early summer, when high temperatures and rainfall conspire to melt the snowpack in the mountains. The area of the Fraser River basin is huge, about a quarter of the area of B.C., thus the river rises slowly from its low winter level as temperatures rise. Fraser River floods develop slowly, often over a period of weeks (top figure, facing page), and the time and height of the crest can be predicted with some accuracy.

Coastal rivers, such as the Chilliwack, Coquihalla, and Squamish, have much smaller basins than the Fraser River and respond more rapidly to local heavy rainfall and snowmelt. Sometimes, these rivers crest in summer, but floods are most likely in the fall when heavy rains fall on an early snowpack (bottom figure, facing page). Winter storms are less likely to cause flooding along rivers because precipitation at high elevations generally falls as snow.

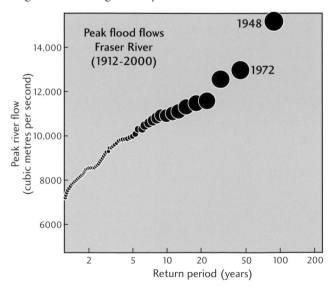

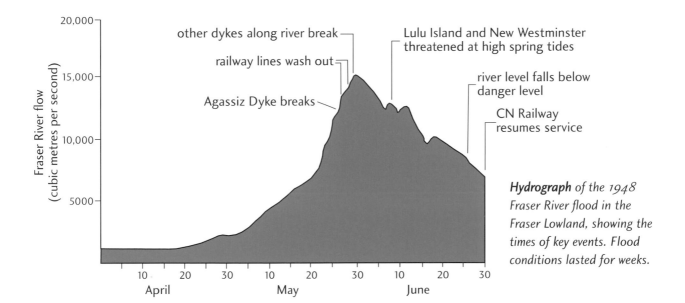

Hydrograph of the 1948 Fraser River flood in the Fraser Lowland, showing the times of key events. Flood conditions lasted for weeks.

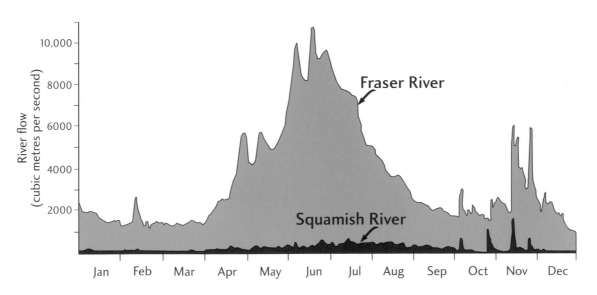

Comparison of hydrographs of the Squamish and Fraser Rivers for 1990. Note that the highest discharge of the Squamish River occurred in the fall during rain storms. The Fraser River also rose dramatically during these storms, but its peak flow was in the late spring when the snowpack melted.

Small streams swell quickly during heavy rains and can flood any time of the year. Such floods are flashy — they strike quickly and generally ebb quickly. Small streams flowing off the North Shore mountains, such as Brothers and Mosquito Creeks, have flooded homes and businesses in North and West Vancouver several times in the twentieth century, most recently during heavy rains in July 1972. The channels of most of these streams have been modified to reduce the possibility of future floods.

Britannia Beach store and gas station in September 1991. A severe rainstorm forced Britannia Creek over its banks and into the community.

Flooding from the sea

Coastal areas may be flooded by the sea when severe storms coincide with high winter tides. In 1948, exceptionally high spring tides broke through some of the coastal dykes protecting the Fraser delta, adding to the damage caused by the great Fraser River flood with which it coincided.

West Vancouver waterfront during a winter storm in January 1999. Severe storms like this can cause considerable damage to property and other infrastructure along the coast.

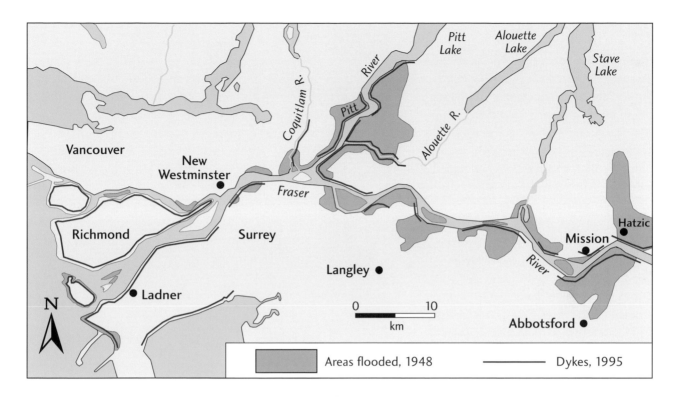

Map of Fraser River dykes and areas flooded in 1948. Most of the flooding occurred in the eastern Fraser Valley. The dykes have been improved to prevent disastrous flooding like that in 1948.

Dykes — severing critical connections

Dykes protect life and property, but at a significant cost to river ecosystems. Naturally functioning rivers, unmodified by dykes, have a myriad of connections to off-channel environments, including back channels, tributary streams, and wetlands that are critical habitat for aquatic plants and animals, and a source of vital nutrients to the river. Dykes disconnect the main river channel from these off-channel habitats, impoverishing the river ecosystem. For example, they prevent salmon from accessing spawning grounds and reduce essential habitat for juvenile salmon living in the river system. Some success has been achieved in restoring these connections by allowing water to flow through dykes at small stream mouths.

The great flood of '48

Over the last 100 years, the Fraser River has risen to flood heights more than twenty-five times (page 68). The greatest flood on record occurred in 1894, when the Fraser Lowland was in the very early stages of settlement and development. The flood caused little damage, but it forewarned of a flood disaster a half century later.

The passage of five decades between the 1894 and 1948 saw the transformation of the lower Fraser Valley into a highly developed agricultural area with growing suburban communities and commercial centres. The Trans-Canada Highway was built through the valley, and Vancouver became an important port at the terminus of two transcontinental rail lines.

Dyke breach near Hatzic during the 1948 Fraser River flood.

In early 1948, river watchers knew that trouble lay ahead when the sudden onset of hot spring weather began to melt the exceptionally thick mountain snowpack. Warm rains melted more snow and contributed to the heavy runoff. In late May and early June, a flood crest rolled down the lower Fraser Valley. Work crews feverishly piled sandbags higher and higher on the dykes, but to no avail — the pressure of the river was too much and the dykes collapsed in more than a dozen places, flooding more than 22,000 hectares of agricultural land, nearly one-third of the entire floodplain area of the lower Fraser Valley. Barns, homes, and animal carcasses floated on the dark swirling waters while boats carried stranded families to higher ground. Sixteen thousand people were evacuated, 2000 homes were damaged, and eighty-two bridges were washed away. Total losses amounted to more than $20 million (about $200 million in today's dollars).

Providing protection from floods

There are several ways to reduce the threat of flooding. Foremost among these are structural measures, including dykes (which can be likened to "corsets of stone"), channel improvements (realignment, dredging, removal of debris, installation of weirs, and bank protection), and diversions. The structural approach is relied on to protect massive development behind the Fraser River in the Fraser Valley.

Dams built for hydroelectric power generation on some tributaries of the Fraser River regulate flows during periods of high runoff and thus reduce the risk of flooding. Reservoir levels are lowest in late winter and spring just before runoff from snowmelt begins. The runoff is stored in the reservoirs and released back to the river system later in the year to provide power and to augment lower fall and winter flows.

In the years after the flood, more than 260 km of dykes were repaired or rebuilt by the Fraser River Dyking Commission. By 1960, some 375 km of river and sea dykes provided protection to most of the vulnerable floodplain and reclaimed tidal lands (page 71). In 1968, the federal and provincial governments established the Fraser River Flood Control Program to further rehabilitate and improve the system of dykes in the Fraser Valley. By 1994, the two governments had spent almost $300 million on flood-control structures and programs. A good investment, most would argue, since a recurrence of a flood similar to that of 1894 could cause at least $2 billion damage. The river dykes successfully withstood the 1972 flood, the largest since 1948, and the sea dykes endured the record high sea levels of 1982.

View south across flooded farmland in the eastern Fraser Lowland during the 1948 flood.

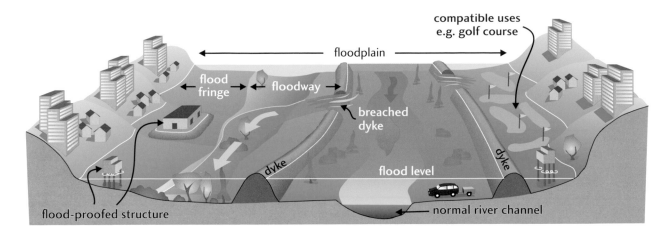

Cross-section of a river valley showing a floodplain, the area that could be inundated during a flood. Flooding is most severe along the floodway. Development on a floodplain is generally discouraged.

Another approach to reducing flood risk is floodplain management. The British Columbia and Canadian governments operate a joint floodplain management program, which strives to ensure that all new development on floodplains will be unaffected by the 200-year flood, the largest flood that can be expected in a 200-year period. The floodplains of more than 120 rivers and streams in B.C., including the Fraser River, have been mapped under this program. Municipal plans, zoning by-laws, building codes, and other regulations are used to discourage residential and industrial development of flood-prone areas. Restrictions are most severe on the part of the floodplain referred to as the **floodway**, where floodwaters are deepest, swiftest, and can cause the greatest damage. Within the **flood fringe**, which is outside and above the floodway, zoning bylaws may require that structures be flood-proofed. Buildings can be engineered to withstand floods or can be placed on gravel platforms above flood level, as has been done in the Pemberton Valley. Nevertheless, critical structures, including hospitals, schools, police stations, and hazardous waste storage facilities, are commonly prohibited in the flood fringe.

Floodplain zoning is arguably a better way of dealing with flooding than structural measures such as dyking. Restricting urban development on lands subject to flooding and dedicating these lands to farming and recreation are less expensive than herculean efforts to hold back floodwaters. Zoning, unlike dyking, allows for natural regeneration of the floodplain, keeping fertile land in food production. If they fail, dykes can actually increase flood losses, for they then function as dams that pond water on the floodplain and impede its subsequent return to the river.

The B.C. Ministry of Environment issues streamflow forecasts during the peak spring flow period. The main purpose of flood forecasting is to provide as much advance warning as possible of an impending flood, but flood forecasts are also used to regulate reservoirs and streams for hydroelectric power, irrigation, domestic uses, and recreation.

House near Pemberton built on a platform of artificial fill to protect it from flooding by the Lillooet River. The low area in the foreground is part of the river's floodplain.

The disastrous 1921 Britannia flood

Driving along the Sea-to-Sky Highway past the Britannia Mining Museum and the old mining town of Britannia, one might not guess that one of the worst natural disasters in western Canada took place here early in the last century.

In its heyday, Britannia was a bustling town at the mouth of Britannia Creek on Howe Sound. When the town was built, no consideration was given to the possibility that Britannia Creek might flood, but that is just what happened on October 28, 1921. The days immediately preceding October 28 were marked by heavy rains, and Britannia Creek ran high. Culverts beneath a large earthen railway embankment that crossed Britannia Creek directly above the town became plugged with debris and a lake began to form. As the lake grew larger, the trapped waters exerted more and more pressure on the embankment. Eventually the dam failed and a torrent of debris-laden water cut a swath through the community. The flood killed thirty-seven people and destroyed about half of the 170 houses in the town. A few large boulders carried by the floodwaters remain in the town, silent reminders of this past catastrophe.

The mining community of Britannia, built on the floodplain of Britannia Creek, before and after the catastrophic flood of October 1921.

Mountain watersheds

Vancouver is blessed with some of the cleanest and safest drinking water in the world. Three magnificent watersheds, the Capilano, Seymour, and Coquitlam River basins, provide water to nearly two million people in eighteen municipalities. The main stream in each of the watersheds has been dammed to create a storage reservoir. Pipelines from the reservoirs distribute water throughout the Greater Vancouver area. The Capilano is the westernmost of the three watersheds. The Seymour watershed is east of the Capilano, and still farther east and separated from the Seymour watershed by Indian Arm is the Coquitlam watershed. Each of the three watersheds has an area of about 200 square kilometres.

All three watersheds receive an astonishing amount of precipitation, up to 500 cm each year, more than four times the amount

Three mountain watersheds supply water to municipalities in the Greater Vancouver Regional District.

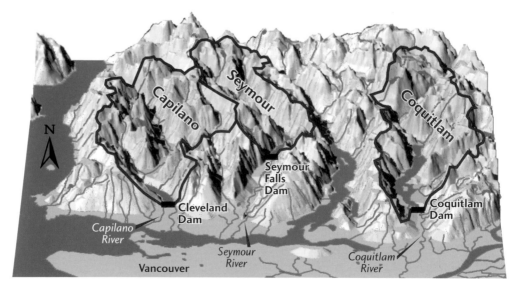

Capilano Lake, one of three reservoirs that provide water to residents of the Greater Vancouver Regional District. The lake fills the lower part of a large valley carved by glaciers. Cleveland Dam (foreground) impounds the lake and straddles the head of a deep, narrow canyon that has been cut into rock by the Capilano River. The road above the right (east) side of the lake follows a powerline that intersects Highway 99 south of Britannia. An upgrade to this road has been suggested as an alternative route to Squamish. This route would avoid most of the rockfall and debris flow problems that plague the Sea-to-Sky Highway, but at what cost to our water supply?

recorded at the Vancouver International Airport. Reservoirs in each of the watersheds store only a small fraction of the winter and spring runoff for the summer months when demand is greatest. Most of the runoff passes through the dams and enters the sea.

Managing our watersheds

The Greater Vancouver Regional District (GVRD) manages the three watersheds. In addition to ensuring an adequate supply of high-quality water, GVRD has secondary concerns: biodiversity, wildlife, esthetics, and cost effectiveness.

Since the first water supply intake was constructed on the Capilano River in 1886, the approach to watershed management has undergone major changes. In the early years, Vancouver's watershed valleys attracted loggers, miners, and homesteaders. Gradually, the GVRD acquired ownership of the private lands and obtained leases over Crown lands, and its present rights in the three watersheds were secured by the end of World War II. The original Crown leases specified that the lands could be used only as a source of water supply, but logging continued on private lands for some time, and a railway extended nearly to the head of the Coquitlam River. Heavy insect damage led to substantial salvage logging and associated road building on both private and lease lands in the 1950s and 1960s. In 1967, the original leases were amended, allowing forest management and commercial logging on a sustained-yield basis. After 1991, sustained logging was phased out and GVRD focused on protection and enhancement of water quality in the watersheds.

Critical issues

Certain issues dominate scientific and public debate over management of the watersheds. These issues include turbidity, acidity, logging and forest practices, roads, public access and recreation, biodiversity, fire, forest pests, long-term climate change, and water quality. Geology plays an important role in some of these issues, notably turbidity and acidity.

Hey, there's mud in my water!

Sometimes, after heavy rains in the fall, our water turns murky with fine sediment and organic particles. Turbidity is a major concern to the public when it shows up in tap water. Although cloudy water itself is not a health risk, fine mineral and organic particles can reduce the ability of chlorine to kill bacteria and other microorganisms. As a consequence, turbid water must be treated more heavily with chlorine, which has an unpleasant odour.

The cause of the turbidity is much debated. Many people believe the problem results from logging and road building in the watersheds. However, several investigations by geotechnical and hydrological consultants have shown that most of the sediment responsible for turbidity originates close to the reservoirs and that logging is not an important factor. The most important sources of turbidity are natural landslides at or near the shores of the reservoirs. Other important factors are erosion of silt-rich Ice Age sediments along the shores of the Capilano and Coquitlam reservoirs by strong waves, and re-mixing of sediment on the reservoir floors. Although some landslides in the watersheds appear to have been caused by logging, they have occurred at considerable distances from the reservoirs and appear to have contributed little or no turbidity to the water supply.

The Britannia Mine — mobile metals and acid waters

The former Britannia Mine, once the largest copper producer in the British Empire, contributed to the growth and status of B.C. as an international centre for exploration and mining. Unfortunately, it has left us with a difficult environmental problem.

Development and closure of the mine predated our present understanding of the damaging effects of acid rock drainage. The Britannia ores contain abundant pyrite, a common iron sulphide mineral also known as "fool's gold" because of its brassy colour. Pyrite reacts with oxygen and water to form sulphuric acid. Over 160 km of tunnels and five pits were created during the operation of the Britannia Mine, exposing a large amount of acid-generating pyrite to air. Rainwater and snowmelt enter the mine through a pit at the top of the mountain. The water reacts with pyrite and other sulphide minerals, leaching metals that are carried from the tunnels to Britannia Creek and Howe Sound. The problem with the water is not the acidity itself, but rather the metals that it transports. The copper content of the drainage water reaches levels that are toxic to aquatic life, including juvenile salmon migrating along the Howe Sound shoreline from the Squamish River.

Prevention of acid rock drainage at the Britannia Mine requires removing water, oxygen, or pyrite from the acid rock drainage equation. This would be extremely difficult and costly because of the vast surface area of pyrite exposed to air and water in the mine tunnels. A more realistic solution, which has been adopted by the provincial government, is to treat the runoff from the mine. Addition of lime to the mine waters neutralizes the acid and precipitates metals in a harmless sludge that can be disposed of safely.

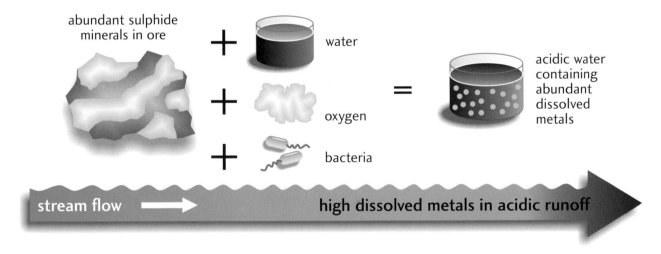

When exposed to water, oxygen, and bacterial action, sulphide minerals such as pyrite produce acidic waters that can carry high concentrations of dissolved metals.

Acidic waters

Vancouver's soft water is naturally acidic (pH = 6.3; neutral water has a pH of 7) and so leaches metals such as copper and lead from the pipes through which it travels. In many areas of the world, the natural acidity of rain is reduced through chemical reactions with minerals in soils and rocks. However, surface waters in the Vancouver area remain acidic for at least three reasons: 1) the watersheds are underlain by granitic and metamorphic rocks that are resistant to chemical attack, 2) the water rapidly runs off steep slopes and thus there is little time for chemical reaction, and 3) high rainfall has leached most local soils of natural calcium and magnesium ions that can neutralize acidity. Existing levels of acidity pose no health problems. The acidity of our water has not changed in recent decades, in spite of marked increases in acid-forming automobile and industrial emissions such as nitrous oxides and sulphur dioxide. These increases are possibly being counter-balanced by parallel increases in acid-neutralizing agricultural emissions such as ammonia produced by livestock operations in the eastern Fraser Lowland.

The subterranean workings of the Britannia Mine generate metal-rich, acidic waters that drain into Howe Sound. The acidic waters float on denser seawater and are lethal to life, including juvenile salmon in nearshore areas.

metal-rich waters open pit rain

Britannia Creek

Highway 99

mill

orebody

mine tunnel

layer of fresh water

metal-rich waters

Howe Sound

volcanic rock

seafloor mud

fan sand and gravel

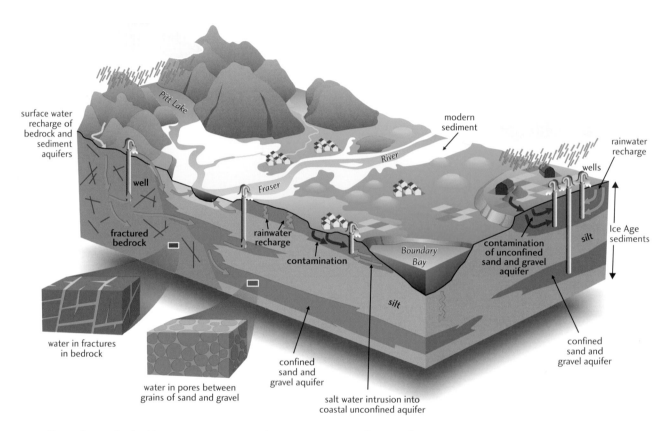

surface water
recharge of
bedrock and
sediment
aquifers

Pitt Lake

modern
sediment

rainwater
recharge

well

wells

Fraser

River

rainwater
recharge

Ice Age
sediments

fractured
bedrock

contamination

Boundary
Bay

contamination
of unconfined
sand and gravel
aquifer

silt

silt

water in fractures
in bedrock

water in pores between
grains of sand and gravel

confined
sand and
gravel aquifer

salt water intrusion into
coastal unconfined aquifer

confined
sand and
gravel aquifer

*Groundwater in the Vancouver area occurs in two main types of materials. Ice Age (orange) and modern
(yellow) sands and gravels, and fractured bedrock (purple). Water is stored in pores between grains in
sediment and in fractures in rock. Unconfined aquifers at the surface are vulnerable to contamination from
human activities. Deeper confined aquifers are more protected from surface contamination by overlying
low-permeability silt.*

Water underground

GROUNDWATER is a hidden, often forgotten resource, yet about one of every four Canadians relies on groundwater. In rural parts of the Fraser Lowland, groundwater supplies much of the total water needs. Domestic consumption accounts for about half of this use, and farming and industrial activities the remainder.

Groundwater is supplied from over 10,000 wells in the Fraser Lowland. Most of these are small wells that supply single households or farms, but some high-yield wells support large communities and industries such as fish hatcheries.

What is groundwater?

Groundwater is subsurface water, part of the global water cycle. Some of the water that falls as rain and snow infiltrates the ground and descends to the water table, where it becomes part of the groundwater system. Streams and rivers also contribute to ground-water.

An aquifer is a body of rock or sediment that can provide useful amounts of groundwater to wells and springs. Aquifers include a wide variety of earth materials, ranging from **porous** and fractured rocks to sand and gravel. In the case of sediments and porous sedimentary rocks, groundwater resides in the spaces, or interstices, between grains (**pore space**). In metamorphic, volcanic, and granitic rocks, groundwater is commonly restricted to fractures and

Groundwater flowing from a fracture in rock.

Underground rivers and lakes?

Many people believe that water wells in the Vancouver area tap into underground rivers or lakes, similar to surface rivers and lakes. This view likely derives from the many spectacular photographs of underground caves formed in limestone, which indeed can host underground streams and small lakes. However, free-flowing subterranean water can only occur in soluble rocks such as limestone, where water flowing in cracks can slowly dissolve large openings. In all other rock types, water resides in tiny spaces, typically a millimetre or less in width. It is more accurate to describe aquifers in the Vancouver area as vast underground sponges than underground rivers or lakes.

faults. An aquifer must not only contain water, but transmit it. Clay has 35 percent or more pore space or **porosity**, and consequently can contain large amounts of water. This water, however, is virtually immovable due to the extremely small size of the pores. Scientists refer to such materials as **impermeable**. The **permeability** of rock is limited by the abundance and interconnection of fractures. Most wells in bedrock produce at rates of 1 to 10 litres per minute, sufficient for a household but not more. In contrast, sand and gravel have high permeability, and water moves rapidly through large interconnected pores between grains. The best sources of groundwater in the Vancouver area are Ice Age and modern sands and gravels. All high-production municipal and industrial wells draw from aquifers in these materials.

Another important aspect of aquifers is their location. Many shallow aquifers are unconfined; that is, they are not covered by impermeable earth materials. An example would be a water-saturated sand and gravel layer lying at the surface, such as occurs in the Abbotsford area (see page 89). **Unconfined aquifers** are tapped by relatively shallow wells and are

volatile hydrocarbon gases

gasoline pool

water table

dissolved contaminant plume

groundwater flow

unsaturated zone

saturated zone

Contamination of groundwater by leakage from a buried gas storage tank. Gasoline is both immiscible in and lighter than water and therefore pools at the top of the water table. Some components of gasoline, however, can dissolve in water and are dispersed in the direction of groundwater flow.

recharged directly from the surface. Consequently, they are highly vulnerable to contamination by domestic, industrial, and agricultural wastes. Most deeper aquifers are covered by impermeable earth materials and thus are confined. Recharge of **confined aquifers** may be indirect — the water may originate as runoff from distant higher ground and may pass through a variety of geological formations before reaching the aquifer. Because confined aquifers have an impermeable cap, they are less susceptible to contamination than near-surface, unconfined aquifers. The impermeable cap can also create a situation where water in the underlying aquifer is under pressure and, when drilled, will vigorously flow out onto the surface as an **artesian well.**

Water can reside in the Earth for periods ranging from days in the case of shallow aquifers to hundreds of thousands of years in some deep confined aquifers. It's hard to believe that the glass of groundwater you drink might be hundreds of years old! In the Vancouver area, shallow groundwater tends to flow from mountains and uplands towards adjacent valleys, where it feeds springs that contribute to streamflow. This water is commonly months to decades old. Deeper-flowing groundwater has much longer residence times and may reach the Strait of Georgia, forming freshwater springs on the seafloor.

As water slowly flows through the Earth, it picks up or leaves behind naturally occurring elements and compounds. The chemical composition of groundwater thus changes slowly over time. As a general rule, the longer that groundwater resides in an aquifer, the more its chemistry

My water comes from Mount Baker!

"I have great well water; it comes from Mount Baker!" Many homeowners on the Gulf Islands and elsewhere, proud of the good taste and abundant supply of their well water, believe they have an underground link to snowmelt on Mount Baker. This legend is incorrect. Studies have shown that groundwater on the Gulf Islands is derived from rainfall on the islands, not from the mainland.

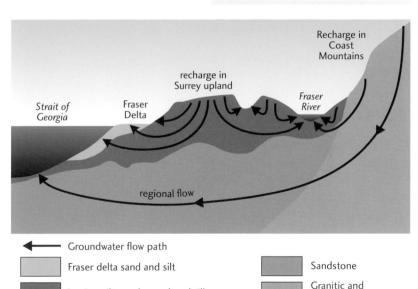

Groundwater flow path

Fraser delta sand and silt

Ice Age silt, sand, gravel, and till

Sandstone

Granitic and metamorphic rocks

Generalized groundwater flow paths beneath the western Fraser Lowland. Groundwater flows from mountains and uplands, where it is recharged by infiltration, to valleys, where it forms springs that feed streams and wetlands.

changes from its original composition. The residence time of water in shallow unconfined aquifers in the Fraser Lowland is brief, and the dissolved mineral content of these waters is correspondingly low except where polluted by petroleum, chemical compounds, or agricultural waste. In contrast, groundwater in fractured bedrock may have high concentrations of dissolved elements and compounds, in part because it has resided there for a long time. Fractures in some granitic and metamorphic rock in southwestern B.C. contain lead, zinc, copper, and arsenic sulphides deposited many tens of millions of years ago. The water in these fractures has slowly dissolved these minerals, substantially increasing its metal and arsenic contents in some areas.

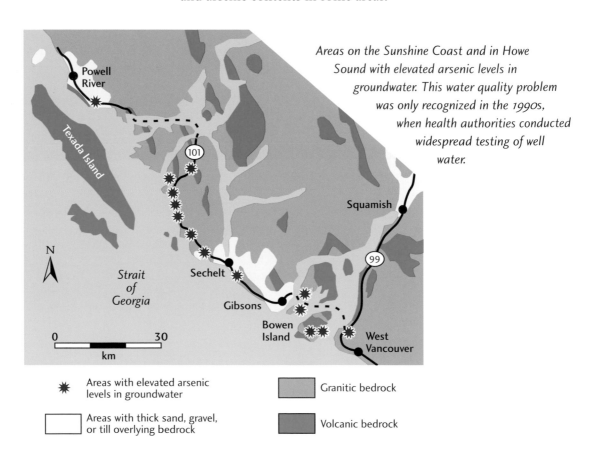

Areas on the Sunshine Coast and in Howe Sound with elevated arsenic levels in groundwater. This water quality problem was only recognized in the 1990s, when health authorities conducted widespread testing of well water.

Areas with elevated arsenic levels in groundwater

Areas with thick sand, gravel, or till overlying bedrock

Granitic bedrock

Volcanic bedrock

Storing natural gas underground?

A sequence of sandstone, conglomerate, and mudstone up to several kilometres thick (the Cretaceous Nanaimo Group and early Tertiary Huntingdon Formation) underlies much of the Fraser Lowland and the adjacent northern Puget Lowland. In most areas, these rocks are buried beneath thick Ice Age and modern sediments (pages 13 and 20), but they are exposed at the surface in Stanley Park, Burnaby Mountain, Bellingham, and many other places.

Schematic diagram illustrating how groundwater becomes contaminated with arsenic. Arsenic-free sediments overlying arsenic-rich rock contain groundwater with low concentrations of the element.

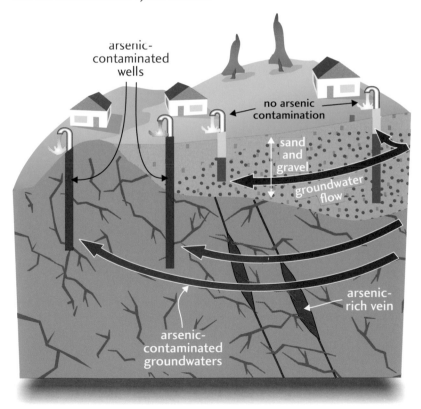

arsenic-contaminated wells

no arsenic contamination

sand and gravel

groundwater flow

arsenic-rich vein

arsenic-contaminated groundwaters

Too much arsenic!

Arsenic is a widely distributed element in the Earth's crust. It generally occurs in small amounts in rocks, but it can become concentrated in groundwater. High arsenic concentrations are one of the most common groundwater problems in rock aquifers worldwide. Long-term exposure to elevated levels of arsenic is known to increase the incidence of some cancers in humans. The Canadian Drinking Water Standards currently set 25 parts per billion as the maximum allowable concentration of arsenic in water, although this level could change in the future as our understanding of what constitutes a "safe level" improves. Testing of water wells over the last decade has identified extensive areas on the Sunshine Coast, on islands in Howe Sound, and in the Fraser Lowland where arsenic levels in groundwater derived from rock aquifers exceed the standard, and alternative water sources must be found. Fortunately, water obtained from shallow wells in sand and gravel in these same areas rarely contain high arsenic levels, as the source of the arsenic is the underlying rock.

Sandstone and conglomerate have much higher porosities than granitic and metamorphic rocks and thus may contain larger amounts of groundwater. However, aquifers in sedimentary rocks in the Fraser Lowland are too deep to be exploited in a cost-effective manner.

In recent years, proposals have been made to use these deep sandstone aquifers as natural gas storage reservoirs. Proponents argue that gas produced from fields in northeastern B.C. and shipped by pipeline to the Lower Mainland could be stored in the rocks and later recovered for use during periods of high demand. This idea has been met with vocal opposition because of concerns that the gas might contaminate near-surface aquifers. Detailed subsurface geological studies would be required before government could approve any such development. It would have to be shown that the subsurface storage reservoirs have a cap of impermeable mudstone and that gas would not migrate to shallow depths along faults or fractures, conditions not likely to be met in the Fraser Lowland.

Water in sand and gravel

Most of the groundwater used in the Vancouver area comes from thick Ice Age sediments, which are a potpourri of different geological materials, ranging from gravel to clay. Perhaps the most striking attribute of the sediments is their complexity. Some subsurface gravel units, for example, grade into sand over short distances, and both may be overlain abruptly by silt and clay. As a consequence, the Ice Age sediment sequence in the Fraser Lowland contains a large number of aquifers that occur at different depths and in different areas. Over 200 aquifers have been identified in the Fraser Lowland, and more are being found all the time.

The Abbotsford, Brookswood, and Hopington aquifers are three of the largest and most heavily used aquifers in the region. Water from these aquifers is used for domestic, agricultural, and industrial purposes. All three aquifers are near-surface groundwater systems,

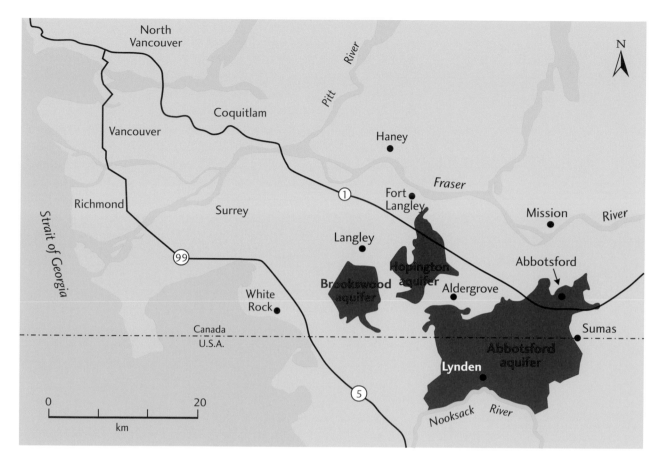

The map contains the following labels:

North Vancouver · Coquitlam · Vancouver · Haney · *Pitt River* · *Fraser* · Fort Langley · Mission · *River* · Richmond · Surrey · *Strait of Georgia* · 99 · Langley · **Hopington aquifer** · Abbotsford · **Brookswood aquifer** · Aldergrove · White Rock · Canada / U.S.A. · Sumas · **Abbotsford aquifer** · **Lynden** · 5 · *Nooksack River* · N

0   20   km

and all three, to some extent, have become contaminated by pollution. The Abbotsford aquifer, located south of Abbotsford, covers an area of about 200 square kilometres in B.C. and Washington. The Brookswood aquifer covers an area of about 50 square kilometres south of Langley. The Hopington aquifer, which is south of Fort Langley, is also about 50 square kilometres in area.

The three aquifers consist of large bodies of sand and gravel deposited by streams flowing off melting glaciers about 13,000 to 12,000 years ago. The Abbotsford aquifer comprises gravelly **outwash** and sandy deltaic deposits, and the Brookswood and Hopington aquifers are large marine deltas that formed along an old sea coast. All three aquifers are recharged from the surface by a

Location and extent of Abbotsford, Brookswood, and Hopington aquifers, three of the largest and most important aquifers in the Fraser Lowland.

Finding underground water is a tricky and expensive business. Shallow "dug" wells are not costly but are only possible where water is close to the surface and is not vulnerable to contamination. Wells usually have to be drilled to depths of several tens of metres, or deeper. The cost of such drilling is high, often many thousands of dollars, and there is no guarantee that sufficient potable water will be found to meet needs.

Enter the "dowser" or "water witch." Dowsers claim to locate subsurface water through the use of hand-held sticks, rods, and other devices. The traditional tool is a forked hazel stick. The dowser loosely holds the two forks of the stick in his hands, with the base projecting outward, parallel to the ground. He or she then walks over the land; the base of the stick swings downward, sometimes suddenly, where water supposedly occurs below the surface.

Dowsers claim that the strong pull on the dowsing rod is due to near-surface electric currents or magnetic fields. Most scientists, however, argue that the currents and fields are much too weak to produce such an effect, and that dowsers are deluded or

combination of direct rainfall, infiltration from streams, and runoff from adjacent higher ground. Groundwater in the Abbotsford aquifer flows mainly to the south, away from higher terrain to the north and west. Groundwater in the Brookswood aquifer flows to the northwest, west, and southwest, away from recharge areas to the east. Water levels in the three aquifers show annual fluctuations of up to 5 m. Levels are highest in February and March at the end of the wet season and lowest in October and November after the summer dry season and before the onset of fall rains. Most wells drilled into these aquifers have good yields. Production rates of up to 30 litres per second are common, and community wells yield as much as 150 litres of water per second.

Spoiling a good thing

Aquifers can be contaminated through careless practices by people who have the attitude "out of sight, out of mind." Groundwater quality in some parts of the Fraser Lowland has been degraded by fertilizers, pesticides, septic field effluent, leaking buried gas tanks, industrial and farm animal wastes, and road salt. Chemicals derived from these materials percolate from the surface down to the water table and then spread outward in the direction the groundwater is flowing (page 82). Over time, a large part of an aquifer may become contaminated, its waters unsuitable for domestic and other uses. Bulk storage of farm animal wastes and their use as fertilizer may be responsible for an increase in nitrates in the Abbotsford aquifer in recent years.

Unconfined aquifers are more vulnerable to contamination than confined aquifers because they are close to the surface and do not have an impermeable cap that impedes infiltration of pollutants. However, not even deep confined aquifers are immune to contamination. Once an aquifer is contaminated, little can be done to correct the problem, at least not in the short term. The source of contamination must be eliminated, and enough time must elapse for natural flow to flush the polluted water from the aquifer. The rate of this natural flushing depends on the permeability and extent of the aquifer and the type of pollutant, but it can take years or even decades for water quality to return to acceptable levels. The cost of the pollution to those dependent on the aquifer for their water supply thus can be very high, a good example of the appropriateness of the maxim "an ounce of prevention is worth a pound of cure."

Another way to ruin an aquifer is to deplete it by extracting too much water. In arid parts of the United States, for example, many aquifers are badly depleted. The water table has fallen, shallow wells have gone dry, groundwater flow that sustains streams has diminished, and drilling and pumping costs required to tap deeper water have risen. This has not yet become a problem in the Fraser Lowland because the amount of water recharged to aquifers exceeds the amount removed. This is true even for the Abbotsford aquifer, which is the most heavily used groundwater system in the region, but it could become a problem in the future.

charlatans. Perhaps some dowsers are charlatans, but most are well meaning and adamantly insist that the scientists are wrong. But controlled tests show that dowsers do no better at locating water than would be expected by chance. Successful dowsers seem to have a basic understanding of where groundwater is likely to occur, based on years of experience in the field. The odds that they will find groundwater are, therefore, higher than those of an ordinary person. However, you tend to hear of the spectacular successes of dowsers, while the failures are never mentioned.

Geologists are often contracted to give advice on where groundwater can be found. They base their recommendations on knowledge of local geology, recharge areas, and groundwater flow. Geologists tend to be more circumspect than dowsers in their recommendations because they know how complex aquifers are and how little is generally known about subsurface geology. However, their recommendations are likely to be more secure if there are already producing wells in the area and if there is abundant subsurface geological information derived, for example, from past drilling.

Hope Slide, the largest historic landslide in western Canada. The top photo was taken on January 11, 1965, two days after the slide. By the time the bottom photo was taken in 1988, vegetation was re-established on the bottom part of the slide.

The ground rules — landslides in the Vancouver area

Without warning late on the night of January 9, 1965, a huge mass of rock, enough to fill B.C. Place stadium several times over, slid away from the flank of Johnston Peak, 17 km southeast of Hope. The rock fragmented as it raced down the slope toward Highway 3 in the valley below. Upon reaching the valley floor, the debris obliterated Outram Lake and drove a wave of muddy sediment 150 m up the opposite valley wall. The muddy debris fell back down the slope and flowed both up and down the valley. When the dust settled, more than 3 km of Highway 3 were buried beneath as much as 80 m of debris and four motorists were dead.

Rockfall on the Trans-Canada Highway in the Fraser Canyon near Yale.

The *trigger* for the "Hope Slide" is unknown. Weather, human activity, and ground-water conditions apparently were not factors. For many years it was thought that two small earthquakes, recorded on a **seismograph** at Penticton at the time of the slide, were the triggers. A recent reexamination of the **seismograms**, however, showed that the inferred earthquake traces were really the result of the land-slide debris striking the valley floor.

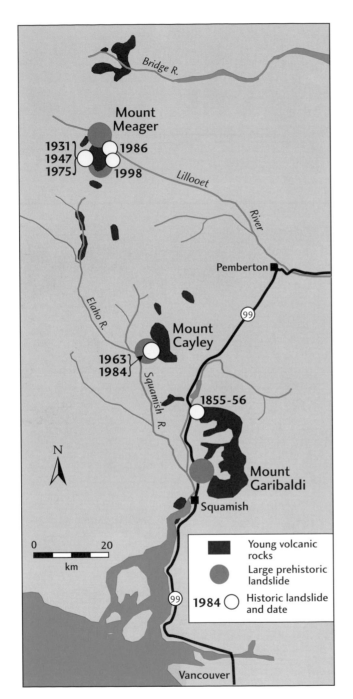

The *cause* of the landslide, however, is known. The metamorphosed volcanic rocks that slid off Johnston Peak are riddled with joints and fractures, some of which dip parallel to the steep mountain slope. In addition, the rupture surface partly coincides with one or more thin, weathered and weak layers of volcanic rock that dip parallel to the slope.

Treacherous slopes

Fortunately, landslides the size of the Hope Slide are rare events. Smaller slope failures, however, are common in the Vancouver area and are capable of considerable damage (see page 93). They have an uneven distribution related to local topography and geology. Most landslides occur on steep slopes and consequently are much more common in the Coast and Cascade Mountains than in the generally flatter Fraser Lowland. However, not all steep slopes are unstable. Rocks that have few fractures and other planes of weakness can stand indefinitely in near-vertical faces, as is evident, for example, at Stawamus Chief near Squamish. Highly faulted, jointed, or bedded rocks, on the other hand, have a higher likelihood of failing, especially where the fractures or beds dip in the direction of the surface slope (facing page).

Young volcanoes in southwestern B.C. are particularly prone to landslides. The slopes of the volcanoes can be very steep and extremely

Locations of large historic and prehistoric landslides on volcanoes in southwestern British Columbia.

unstable. A disproportionate share of the large historical landslides in southwestern B.C. have occurred on these slopes, namely at Mount Cayley, Mount Meager, and near Mount Garibaldi.

Most of the steep escarpments bordering uplands in the Fraser Valley are formed of loose Ice Age sediments. Homes at the edges of these escarpments are highly desired for the views they command. Houses have also been built at the base of many of the escarpments without consideration for the stability of the slope above. Some of the escarpments have failed during periods of heavy rain, damaging and destroying property, and similar failures can be expected in the future.

Fractured granite along Highway 99 at Porteau Cove. The rock slope was undercut during construction of the highway in the 1950s. Conspicuous fractures dip parallel to the slope and are planes of weakness along which the rock may slide, as shown in the schematic drawing at the upper right.

95

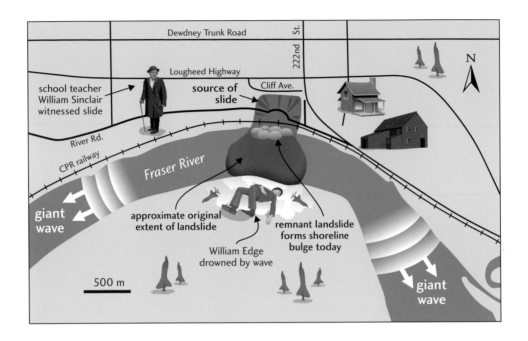

The diagram labels, reading across the map:

Dewdney Trunk Road

222nd St.

Lougheed Highway

Cliff Ave.

N

school teacher William Sinclair witnessed slide

source of slide

River Rd.

CPR railway

Fraser River

giant wave

approximate original extent of landslide

remnant landslide forms shoreline bulge today

William Edge drowned by wave

500 m

giant wave

The Haney landslide — giant waves on the Fraser

At about 3:30 pm on January 30, 1880, 8 hectares of land on the north bank of the Fraser River at Haney suddenly gave way and slid into the river. The movement was rapid and, according to eye-witnesses, was accompanied by a cannon-like roar. Part of the slide mass moved nearly one kilometre across the Fraser River and temporarily blocked its flow. The rapid movement of land into the river produced a displacement wave, estimated at the time to have been about 20 m high. This wave "leapt the bank and mowed down 15 acres of trees as though they were ferns. Farther back, giant fir trees were stripped of their branches 20 feet from the ground" [*The News*, January 6, 1982]. William Edge, a farmer, was swept away by the wave and left hanging in one of his orchard trees. Waves also travelled several tens of kilometres both up and down the Fraser River; 15 km upstream from the slide, the main wave was still 3 m high. These waves destroyed the public wharf and fishery building at Haney and swamped and sank most boats on the Fraser River for a

The Haney landslide occurred without warning on January 30, 1880, when a river bank of glacial-marine clay, which underlies much of the Haney-Maple Ridge area, collapsed into the Fraser River. Waves, 20 m high, moved up and down the river from the site of the landslide.

Cause versus trigger

Geologists distinguish *causes* and *triggers* of landslides. The distinction between these two terms is more than one of semantics. The cause of a landslide is the combination of geological, topographic, groundwater, and human factors that make a slope vulnerable to failure. All other things being equal, steep slopes are more likely to fail than gentle slopes. And steep slopes developed in rocks with beds or joints that dip in the direction of the slope are more likely to fail than slopes formed of structureless rock. Some types of earth materials, for example **Quaternary** volcanic rocks and glacial-lake silt and clay, have low strengths and are prone to failure. Water-saturated sediments and rocks are more likely to slide than dry materials. The pressure of water in pores and fractures may rise, lowering the strength of the material and possibly inducing failure. Finally, humans can modify slopes and cause them to fail. They may, for example, undercut a slope or load its top, both of which increase stress on the slope materials. Blasting during road construction commonly fractures rock and may create unstable conditions. Most **rockfalls** along the Sea-to-Sky Highway between Horseshoe Bay and Britannia are the result of highway construction during the 1950s.

The trigger of a landslide is the event that sends the slope "over the edge," causing it to fail. To use another metaphor, it is the "straw that breaks the camel's back." An earthquake, heavy rainstorm, freeze-thaw cycle, or some other event may trigger a landslide. The trigger may be a very ordinary event, one so ordinary that it may be difficult to detect. The trigger for the huge Hope Slide, for example, is unknown. It is likely that many slopes slowly deteriorate over periods of centuries or millennia to a critical threshold, at which time a trivial event triggers failure.

distance of 30 km from the slide. One boat was carried 100 m inland and left on a hill 12 m above the river.

The landslide occurred in thick silty and clayey marine sediments deposited at the end of the Ice Age about 13,000 years ago. Newspaper reports at the time of the landslide mention a layer of sand or gravel within the clay that was discharging water onto the river bank. It is possible that high water pressures developed in this layer and caused the overlying sediments to fail. The landslide also occurred at the outside of a bend in the river channel where bank erosion would be greatest. No earthquakes or abnormal weather were reported at the time of the landslide, nor was the failure caused by human activity.

The Haney slide appears to be an unusual event because no other landslides of this type and size have occurred in the Fraser Lowland in historical time. A repeat of a Haney-type slide in the western Fraser Lowland could be disastrous now, because the area is heavily urbanized and displacement waves would cause extensive damage to development along the Fraser River.

Rockslides — the Jane Camp disaster

The Jane Camp slide, which occurred on March 22, 1915, near the mining community of Britannia, is a classic example of a **rockslide**, a common type of landslide involving a sliding type of movement. Although few people have heard of the Jane Camp slide, it ranks as B.C.'s worst natural disaster in terms of loss of human life.

Jane Camp was a small community located near the top of Britannia Mountain adjacent to underground tunnels of the Britannia Mine. The camp was rudimentary, typical of the mining communities of the late 1800s and early 1900s, but it housed a large number of miners, women, and children. Shortly after midnight on March 22, 1915, about 100,000 cubic metres of rock and snow slid away from the steep mountainside above Jane Camp and ploughed into bunkhouses and family homes where fifty-six people perished. Residents received no warning of the impending landslide, but a gaping tension crack on the ridge crest above the camp was observed and photographed some time before the disaster. Sadly, company geologists inspected the mountainside two days before the landslide and stated that it was "solid." The cause of the landslide is uncertain, but use of explosives in underground tunnelling in fractured metamorphosed volcanic rocks likely contributed to a progressive deterioration of the ridge that ultimately failed. The slide was preceded by a thaw, leading to speculation that infiltration of water into the slope triggered the failure.

Jane Camp rockslide of March 22, 1915. The landslide claimed the lives of fifty-six men, women, and children who were living in a makeshift mining camp. The photo was taken from the mining camp (foreground) and is a view upslope towards the source area of the landslide, which is obscured by the rock face in the upper right. The width of view at the bottom of the photo is about 150 m. Vancouver Public Library, Special Collections, VPL 13896.

Rockfalls

Rockfall is the rolling or bouncing of large fragments of rock down steep slopes (pages 93 and 95). A rockfall may involve only a single large block or boulder. More commonly, however, a large number of blocks collide with one another and with the ground as they bound down the slope. Intensely fractured or jointed rocks are most susceptible to rockfall, and triggering mechanisms include freeze-thaw activity, heavy rains, and earthquakes.

Rockfall is the process responsible for talus cones and aprons, features that are common in the mountains of western Canada. Spectacular talus aprons can be seen along the Sea-to-Sky Highway north of Alice Lake. Occasionally, large blocks roll beyond the foot of the talus slope into developed areas. No such disasters have occurred in the Vancouver area in historical time, but the presence of huge boulders from ancient rockfalls in built-up areas near Hope and Lions Bay indicates that the hazard cannot be ignored.

Rockfalls are common on excavated rock cuts and adjacent, steep, natural rock slopes along highways and rail lines. They pose a major hazard to road and rail traffic in the Fraser Canyon and the Sea-to-Sky corridor and have caused several deaths and numerous interruptions in traffic in these areas. They are most common in the spring when temperatures fluctuate around the freezing point, and in autumn during heavy rains. Many excavated rock cuts are more prone than natural slopes to rockfall because they are very steep and because blasting has fractured and weakened the rock.

Maintenance of rockfall-prone slopes involves considerable effort and expense. Most of the traffic delays we experience along the Sea-to-Sky Highway between Horseshoe Bay and Britannia result from the removal of loose rock from unstable rock cuts, a process referred to as **scaling**. Other measures used to prevent or minimize rockfall include draping threatening rock faces with heavy coarse metal mesh, spraying them with **shotcrete**, bolting fractured rock with long metal rods, and drilling holes in the rock to allow water to drain out more freely and thus prevent the buildup of water pressure. Take a look at the steep, blasted rock slopes between Horseshoe Bay and Britannia next time you drive the highway, and you will see examples of all these defensive measures. Such efforts can reduce, but never completely eliminate rockfalls along the Sea-to-Sky Highway, the Trans-Canada Highway in the Fraser Canyon, or anywhere else where steep slopes of highly broken rock loom above roads and rail lines.

Rockfall is the most common type of landslide caused by earthquakes. A magnitude 7.3 earthquake in 1946 triggered hundreds of rockfalls in the mountains on central Vancouver Island, and an earthquake of similar or larger size in 1874 caused numerous landslides in northern Washington and southern B.C. A future strong

Rockfall defensive measures along Highway 99 near Porteau Cove: rock bolts on a grouted slope, and a metal screen.

earthquake near Vancouver might trigger rockfalls and rockslides along transportation corridors, severing the economic arteries of the city and impeding recovery efforts.

Rockfalls and rockslides into fjords pose a hazard to some coastal communities in B.C. These landslides, although uncommon, produce large waves (a type of **tsunami**) that can travel tens of kilometres with little energy loss and surge into low-lying communities at fjord heads.

Debris flows

The disaster began innocently enough along the channel of upper Alberta Creek, high above Howe Sound, during an intense rainstorm on February 11, 1983. Water from rain and melting snow pooled behind a mass of logs and sediment in the channel. Soon, the force of the water caused the debris dam to fail, and a slurry of mud, boulders, and logs began to flow down the channel. The **debris flow** grew in size as it scoured additional debris while moving down the channel; it probably had a volume of about 5000 to 10,000 cubic metres when it reached Lions Bay. It was here that things turned ugly. The first surge of the flow, carrying tree trunks and boulders the size of cars and sounding like a locomotive, swept through the community and into Howe Sound. It had a peak velocity of about 9 m per second, or 32 km an hour, roughly the same speed at which a world-class sprinter runs. Five surges later, a swath of destruction had been cut through the community of Lions Bay. Although much of the flow was confined to the stream channel, tongues of debris overtopped the channel and smashed into nearby homes. Several houses were damaged or destroyed, and two teenage boys were killed when the trailer in which they were sleeping was overwhelmed by debris. All culverts and bridges, except the BC Rail bridge, were destroyed.

The Alberta Creek disaster is one of several damaging debris flows that occurred along the eastern shore of Howe Sound between Horseshoe Bay and Britannia Beach in the years following

Aftermath of a debris flow that swept down Alberta Creek and through Lions Bay in February 1983, killing two people.

construction of the Sea-to-Sky highway and BC Rail line in the 1950s. The flows destroyed road and rail bridges that spanned several streams, as well as houses built on debris fans at Charles (Strachan) Creek and Lions Bay. At first glance, it might seem that the recent debris flows are unusual and related in some way to human activity, for example logging or development. In reality, recent disastrous flows like those at Alberta Creek in 1983 and at nearby M Creek in 1981 are natural events, part of a sequence of debris flows that have been occurring along Howe Sound since the end of the Ice Age. The fan on which Lions Bay is built is the product of thousands of debris flows similar to the 1983 Alberta Creek event.

You might wonder why people would put themselves at risk by building and living at such hazardous sites. The simple answer is that they were unaware of the risk. In the 1950s and 1960s, few people in B.C. even knew what a debris flow was — the early destructive debris flows along Howe Sound were dismissed as "floods." In such a situation, it is only natural that communities would be located on fans, which are the flattest and most easily developed land on the rugged, steep, eastern side of Howe Sound.

Anatomy of a debris flow. Debris flows along Howe Sound occur on steep slopes and are triggered by heavy rainfall. The fan, most of which is below sea level, was built up over thousands of years by repeated debris flows. The sequence of events is:

1. *Torrential rains swell streams near the mountain crest.*
2. *Sediment slumps into a raging stream, triggering a debris flow that surges down the channel.*
3. *The debris flow swells in volume as it picks up additional sediment and trees.*
4. *The debris flow emerges from the canyon onto a fan where it damages houses, roads, bridges, and a rail line.*

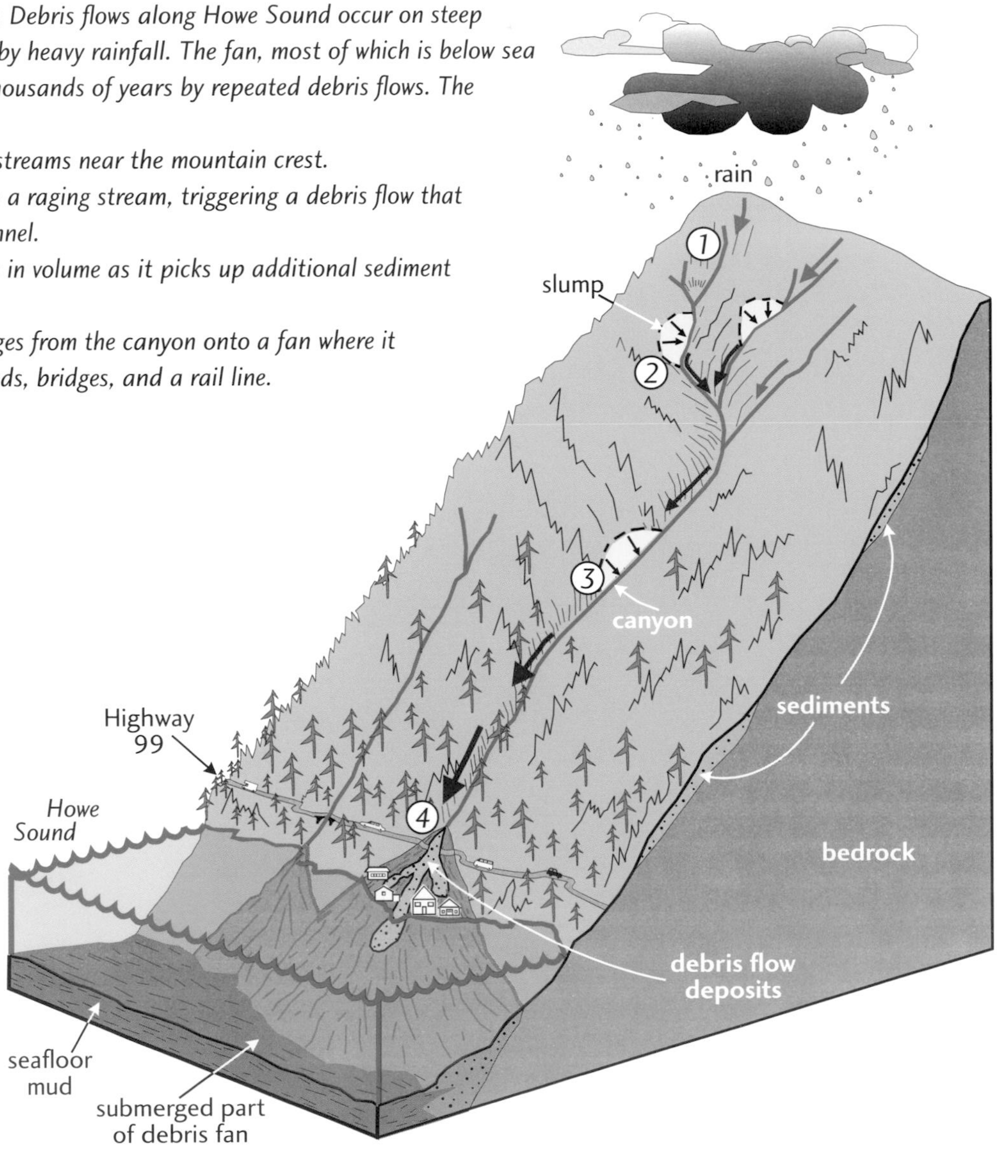

rain

slump

①

②

③

canyon

sediments

Highway
99

Howe
Sound

④

bedrock

debris flow
deposits

seafloor
mud

submerged part
of debris fan

Later, after several debris flow disasters here and elsewhere in western Canada and after scientists and engineers became aware of the European and Japanese experience with this phenomenon, the full scope of the problem along Howe Sound was recognized.

We now know that debris flows are common natural events, not only along Howe Sound, but on most steep mountain slopes in areas of heavy rainfall. They begin with mobilization of debris within a steep, rain-swollen stream channel, as at Alberta Creek, or with a small slump or slide into the channel. In the latter case, the triggering landslide is generally small and involves a shallow layer of weathered sediment. Once started, a debris flow accelerates rapidly and increases in size as it moves down slope. Debris flows generally follow ravines and gullies, but some occur on open slopes and are not channellized (the latter are often referred to as **debris avalanches**). Channellized debris flows spill onto fans or cones, where most of the damage and loss of life occurs.

The M Creek and Alberta Creek disasters in the early 1980s were catalysts for a program of highway improvements and other measures aimed at reducing the debris flow hazard along Howe Sound. The B.C. Ministry of Transportation has spent large sums of money protecting Highway 99, the BC Rail line, and residents in Lions Bay and other, smaller communities in the area. All the old, wood-trestle highway bridges over streams like M Creek were replaced with modern steel-concrete structures with enough clearance to allow the passage of debris flows. Two large debris-flow retention basins were constructed, one at Harvey Creek above Lions Bay and another at Charles Creek. The idea is that the basin dam allows normal streamflows to pass but blocks debris flows. Roads above the dams provide access to the basins so that workers can remove trapped debris flow deposits. The channels below the dams are lined with concrete to prevent erosion and to guide high streamflows directly into Howe Sound. The channel of Alberta Creek was straightened and lined with concrete to carry future debris flows through Lions Bay to Howe Sound, and improvements were made to several other stream channels. No damage has occurred from debris flows along Howe Sound since completion of these works in the mid-1980s at a cost of about $20,000,000.

Underwater landslides

Not all landslides occur on land — steep slopes in lakes and the sea may also fail. These landslides commonly go unnoticed, but they can be damaging to shore installations. A submarine slump adjacent to the Woodfibre pulp mill in Howe Sound in 1957 damaged the mill's shore facilities, and a large landslide on the Fraser delta slope in 1983 came close to destroying a staffed Coast Guard station at the mouth of the Fraser River near Sand Heads.

Underwater landslides are especially common at the fronts of active deltas, such as the Fraser and Squamish deltas. The Fraser and Squamish Rivers carry large loads of sediment that are deposited where the rivers enter the sea. Over time, sediment builds up at the delta front and then slides or flows into the deeper waters of the Strait of Georgia and Howe Sound. Earthquakes or extreme

Learning to live with mountains

Unlike the Swiss or Japanese, Canadians have not lived in mountains for long. Europeans settled southwestern B.C. only 150 years ago, and we continue to learn how mountains work. Unfortunately, nature often teaches us through disasters. The 1915 Jane Camp slide taught us about the danger of steep slopes, and the 1921 Britannia flood provided a lesson about the dangers of living along mountain streams. It is humbling to realize that as recently as the 1970s, highway engineers did not understand the danger of debris flows and built highway bridges designed only for floods. As we have yet to experience the impact on our mountains of a large earthquake near Vancouver, there will no doubt be more hard lessons!

Debris flow dam on Harvey Creek above Lions Bay. The dam allows the passage of normal stream flow but impounds debris flows, thus protecting the residents of Lions Bay.

Debris flow scars on a steep logged slope in the Cascade Mountains near Wahleach Lake southeast of Chilliwack. Several debris flows moved down this slope during a heavy rainstorm. They were caused by the failure of embankment materials along a logging road.

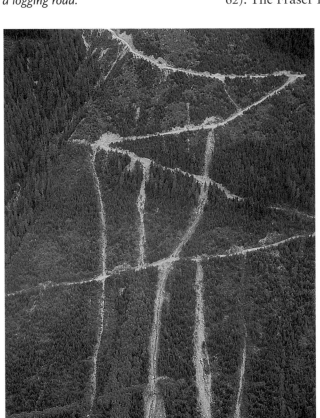

tides may trigger these failures, although the progressive buildup of sediments alone will generally do the job.

The Fraser delta front could fail again at any time, and almost certainly would fail during a strong earthquake. A large submarine landslide could damage the Roberts Bank Deltaport, the Tsawwassen Ferry Terminal, or submarine cables that supply much of the electrical power to Vancouver Island. Failures are most likely, however, off the mouth of the Main Channel at Sand Heads (page 62). The Fraser River supplies an average of 17 million cubic metres of sediment annually to the delta's **distributary channels** below New Westminster. Approximately 80 percent of this sediment is carried in the Main Channel. Today, large amounts of sand are dredged from the Main Channel to maintain a navigable channel to New Westminster. The sediments that do reach the river mouth periodically fail and slide or flow down a deep submarine channel west of Sand Heads to water depths of about 300 m in the southern Strait of Georgia.

Prior to European settlement, the distributary channels of the Fraser River regularly shifted their positions on the delta plain. A metaphor is a loose fire hose spraying water in different directions. The locus of sediment deposition at the delta front shifted with the channels, and sediment accumulated and failed at different places. The "fire hose" today is firmly fixed in one place by river dykes, and virtually no sand reaches the delta slope north and south of the Main Channel. Therefore, the delta is no longer building up evenly over its front.

Forest harvesting and landslides

A debate rages over whether or not forest harvesting increases the frequency of debris flows and other landslides. Clear-cut logging removes the forest cover and increases water runoff. Heightened runoff leads to higher peak stream discharges and, in some cases, to an increase in pore water pressure in soils, which is a common landslide trigger. Clear-cut logging also directly increases water pressure in soils because trees pump large amounts of soil water into the atmosphere through transpiration. The roots of cut trees, however, continue to stabilize the soil mass. Researchers have noted that the root mass decays gradually; many years may pass after logging before landslide activity increases.

The primary culprit, however, is not tree removal as such, but rather improper construction of logging roads. Most landslides on clear-cut slopes originate at poorly engineered or constructed road fills. Roads intercept shallow groundwater, increasing surface runoff in ditches. Failures are especially common where surface drainage is blocked or concentrated, for example, when culverts become plugged with sediment.

The Campus Canyon washout

Short steep gullies at the edges of Ice Age uplands in the Lower Mainland may be catastrophically eroded by uncontrolled runoff during periods of heavy rain. Rain-swollen streams cascading down these normally dry gullies rapidly cut downward and headward to form spectacular canyons bounded by steep walls. The canyon walls collapse and retrogress as the erosion progresses.

The most spectacular example of such an event occurred on the campus of the University of British Columbia in January 1935 after two days of torrential rain following a week of heavy snowfall. Water backed up over a large area of the university campus and overflowed to the north towards the Point Grey sea cliff. Over a period of two days, a raging torrent in a formerly minor gully removed about 100,000 cubic metres of loose Quadra Sand, creating a badland canyon of impressive proportions (photo, next page). The steep walls of the canyon repeatedly collapsed, sending surges of sand and water down to the sea where a large fan formed. As a result of this disaster, proper drains were installed on the UBC campus, including an innovative, vertical spiral drain that drops runoff water from the top of the Point Grey cliff to the sea. Catastrophic gully erosion and bank collapse similar to that at UBC occurred in Coquitlam Valley during the winter of 1951–1952, when about 300,000 cubic metres of sediment accumulated on the valley floor, blocking the Coquitlam River for several days.

Campus Canyon washout, January 1935. Uncontrolled drainage from the UBC campus over a two-day period during torrential rains carved this canyon in loose Ice Age sand, which underlies Point Grey. Photo courtesy of the University of British Columbia Archives.

Avalanche!

OUR COASTAL mountains receive large amounts of snowfall during winter. When snow piles up on steep slopes, it may fail, triggering an avalanche. Avalanches are fast and powerful; some move as fast as an automobile and destroy tracts of forest. A large avalanche may cross a valley and run some distance up the opposite slope. Avalanches regularly disrupt road and rail traffic in some mountain passes and valleys, such as Rogers Pass and the Fraser Canyon. And almost every year, skiers and snowmobilers are killed by avalanches. The toll during the winter of 1997–1998 was particularly high; seventeen people died in avalanches in B.C. and Alberta.

The where and when of avalanches

Avalanches take place on slopes ranging from 20 to 60 degrees, but are most common on moderate (30 to 40 degree) slopes where snow can accumulate to considerable thickness and then suddenly fail. Avalanches involve either loose surface snow or a slab of cohesive snow. They are triggered by rain, snowfall, wind, a change in temperature, or a sudden shock, such as an earthquake or simply a skier or snowmobiler crossing a slope.

Avalanches occur in ravines and gullies and on open slopes. Areas frequently swept by avalanches on forested slopes are easily identified by the absence of trees (photo, next page). In contrast, avalanche areas above treeline are more difficult to recognize.

Avalanches are more common under certain weather conditions, notably during thaws or periods of rain after heavy snowfalls, and when thick dry snow accumulates on old icy snow surfaces. In the

Avalanche tracks and a snowshed along the Coquihalla Highway southwest of Coquihalla Pass and the toll booth. The snow-covered strips above the highway are raked by avalanches in winter. The dark areas of large coniferous trees have been free of avalanches for at least the life of the trees.

latter instance, the buried icy surface is a plane of weakness along which the dry snow easily slides. Winter weather conditions in the mountains of B.C. are closely monitored, and warnings issued when the avalanche hazard is high.

Avalanches and the Coquihalla Highway

The Coquihalla Highway, completed in 1986, crosses the Cascade Mountains between Hope and Merritt. Southwest of the summit, the highway follows a valley with over seventy avalanche paths. On average, 100 avalanches per year occur along this stretch of the highway. Most of the avalanches are small and pose no threat to motorists, but some reach the highway.

The Ministry of Transportation has an extensive avalanche forecasting and control program along this stretch of the Coquihalla Highway. Several weather stations operate throughout the winter, the stability of the snowpack is assessed, and avalanche forecasts are issued at least once a day during high hazard periods. The highway is closed when the hazard is extreme, but this is rarely required because the road has been built in the centre of the valley and is elevated well above the valley floor. **Static defences** provide further protection — dams deflect avalanches, and basins and mounds slow or stop them. A 281-metre-long snowshed was built at a cost of $12 million at a particularly active site to allow avalanches to pass over the highway. **Active defences** involve the use of artillery or explosives to bring down avalanches under controlled conditions. You can see the gun platforms along the highway.

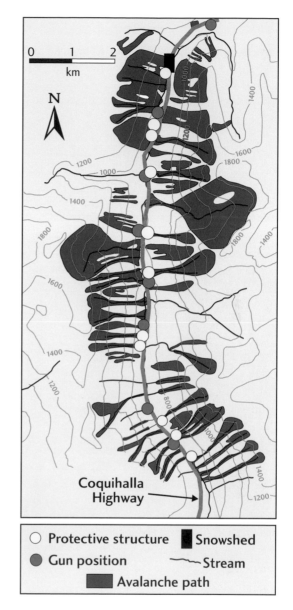

Avalanche paths, shown in pink, along a section of Coquihalla Highway in the steep mountain valley south of the summit. The green lines are contours, which are lines of equal elevation. In this case, the vertical spacing between contours is 200 m.

111

Avalanches in the Fraser Canyon — the winter of '96

Landslide at Conrad, south of Lytton, March 1997. The landslide severed the CN rail line, causing a freight train to derail with the loss of two lives.

Residents of Victoria had never seen anything like it. It began to snow on December 27, 1996, and within four days, about a metre had fallen, paralyzing the city. Vancouver fared a little better but also received a heavy dump of snow. Conditions in the eastern Fraser Valley and Fraser Canyon were abysmal. Heavy snowfall and strong winds created whiteout conditions. Motorists were stranded and were rescued and housed in shelters and hotels. Motorists attempting to drive through the Fraser Canyon took refuge in Lytton and Boston Bar.

The storm left unusually deep snows on the steep walls of the canyon. The snowpack failed in hundreds of places, producing avalanches that swept down onto the Trans-Canada Highway and buried it to depths of many metres. When the snow stopped falling in early January, the Ministry of Transportation faced the daunting task of clearing the many avalanches that blocked the highway between Yale and Boston Bar. Working around the clock, crews managed to re-open the road in a few days, enabling stranded travellers to continue on their way.

A postscript to this story is that rapid melting of the heavy snowpack in the Fraser Canyon in March 1997 triggered numerous landslides, including one at Conrad, south of Lytton, that severed the Canadian National Railway mainline. A freight train plunged into the hole left by the Conrad landslide, killing two engineers, destroying millions of dollars of rail stock, and interrupting rail service for more than a week. The landslide resulted from rapid infiltration of runoff water into sediment fills on which the rail line and highway had been built. The buildup of water increased pressures in the fills, causing them to fail. Proper surface drainage at the site might have prevented this disaster.

THE **Province**

Thursday, March 1, 2001   Vancouver, B.C   www.theprovince.com

60¢

Coin box: 75¢
Outside Lower
Mainland:
$1 minimum

EIGHT-PAGE SPECIAL REPORT STARTS ON A2

QUAKE!

Powerful earthquake causes billions of dollars in damage to Seattle, and scares southern B.C.

Kristi Heim reacts to a crushed car and a collapsed wall at the Starbucks corporate headquarters building in Seattle yesterday.

— Reuters

The Nisqually earthquake — a wakeup call for residents of Vancouver.

On shaky ground — living with earthquakes

W E LIVE IN earthquake country, and it is possible that we may experience a damaging quake at some time in our lives. Big quakes in recent years in Los Angeles, San Francisco, Seattle, and Alaska have alerted us to the damage that can occur when the ground shakes violently. About one earthquake occurs somewhere in B.C. every day, although most are too small to be felt. We can expect an earthquake of the size of the catastrophic 1995 quake at Kobe, Japan, however, on average once every 30 years somewhere in southwestern B.C. or northwestern Washington. So let's take a closer look at these powerful forces of nature.

An earthquake is produced by the sudden displacement of rocks along a fault. The rupture of the rock sends out shock waves that travel outward from the source.

What causes earthquakes?

An earthquake is the result of a sudden release of energy when rocks break and move rapidly past one another along a fault. The sudden slip, which may or may not extend to the surface, causes the ground to shake. The location on a fault where slippage first occurs is called the **focus**, and the position directly above it on the Earth's surface is termed the **epicentre**.

source

The **Richter scale**, devised in 1935 by the renowned **seismologist** Charles Richter, is a numerical scale of earthquake magnitude or strength. Magnitude on the Richter scale provides a measure of the amount of energy released during an earthquake and is determined by the largest amplitude of the shear wave, measured at a specific distance from the epicentre on a seismograph, an instrument that records earthquake waves (page 118).

The amplitude is converted to a Richter magnitude using logarithms. A magnitude 7 earthquake, for example, produces a displacement on a seismograph ten times larger than an earthquake of magnitude 6. However, the energy released, which is proportional to magnitude, is about 30 times greater. To look at it another way, a magnitude 7 earthquake releases about 900 times (30 times 30) the energy of a magnitude 5 quake. Although there is no upper limit to the Richter scale, the strength of

Shaking

Earthquake shaking results from **seismic waves** of a variety of sizes, frequencies, and velocities. As seismic waves move from one kind of material into another, their paths can be refracted or reflected, much as light can be refracted or reflected when passing through water or glass. In addition, two separate waves of the same shape and size may amplify each other, producing a larger and more powerful wave than either of the originals.

There are two basic types of earthquake waves: body waves that travel through the Earth, and surface waves that travel along the Earth's surface. Body waves include **primary** and **secondary waves**. Primary (P) waves are propagated by alternating compression and expansion of material in the direction of movement — a push-pull type of motion. They are the fastest of the seismic waves and the ones that carry sound. Secondary (S) waves involve a shearing motion of material, with oscillations perpendicular to the direction of propagation. The up-and-down and side-to-side oscillations are analogous to the snapping of a rope. Some P and S waves transform into surface waves when they encounter the Earth's surface. One type of surface wave, known as **Love waves**, moves transversely, whipping the ground from side to side without displacing it vertically. Another, called **Rayleigh waves**, has a rotating, up-and-down motion like waves at sea. Love and Rayleigh waves have lower frequencies than the P and S waves that spawn them and usually travel more slowly. However, they may travel great distances, sometimes circling the globe several times before dissipating.

Ground shaking during a strong earthquake may result from a combination of primary, secondary, Love, and Rayleigh waves, and all four wave forms are capable of damage. Earthquake shaking can be likened to a chaotic symphony comprising a number of

Facing page: Earthquake waves are of four principal types with different motions. Primary waves are push-pull; shear waves are like snapping a rope; Love waves whip the ground from side to side; and Rayleigh waves roll the ground like waves at sea. The large arrows indicate the direction of travel of the waves.

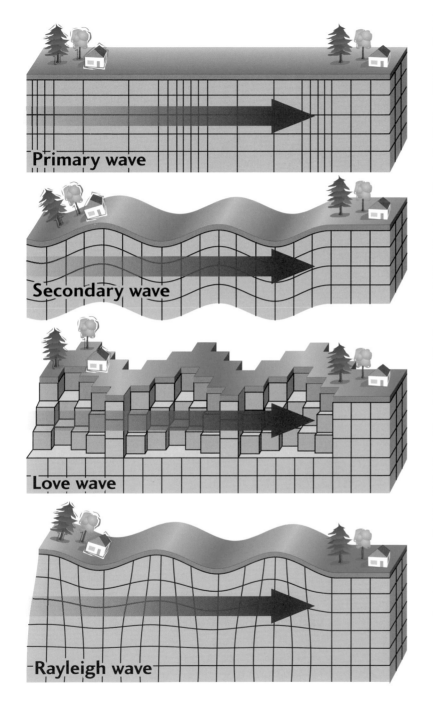

Primary wave

Secondary wave

Love wave

Rayleigh wave

Earth materials produces an actual upper limit of about 9.

Richter magnitude can be related to the damage expected from an earthquake. An earthquake of magnitude 2, for example, is unlikely to be felt, whereas one of magnitude 5 can cause moderate damage very near the epicentre, and an event of magnitude 7 is a major earthquake capable of causing widespread damage.

Recently, there has been a move to change from the Richter scale to the **moment magnitude** scale. The latter scale is based on the seismic moment, defined as the product of the average amount of slip on the fault that produced the earthquake, the area that actually ruptured, and the strength of the rocks that failed. The moment magnitude scale has a sounder physical basis and is applicable over a wider range of ground motions than the Richter scale. Consequently it is increasingly being favoured in reporting earthquake statistics. The scale, however, provides numbers that are close to Richter magnitudes in most cases.

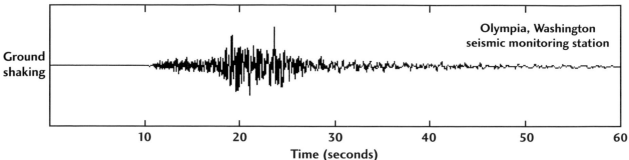

Time (seconds)

Seismogram of the February 2001 Nisqually earthquake centred near Olympia, Washington. The earthquake had a magnitude of 6.8 and caused about 2 billion dollars in damage. The amplitude of the squiggles on the graph is proportional to the intensity of the shaking.

dissonant instruments. However, one type of wave may be responsible for much of the damage, and different types of structures may be preferentially damaged by one type of wave.

Plates and quakes

Stress that causes earthquakes results from the movement of rigid plates that make up the outer shell of the Earth. As noted previously, the plates "float" on the outer mantle, a dense, hot, plastic layer. The plates move at a slow speed, typically several centimetres per year, or about the rate at which fingernails grow. Consequently, we do not notice the movements in our everyday lives. Over time, however, these small movements can build up enough stress to produce large earthquakes.

The west coast of B.C. is located within a belt of earthquakes and volcanoes that encircles the Pacific Ocean, known as the **Pacific Ring of Fire**. This belt marks the boundaries of several plates. Three plates meet off our coast, making the region the most active earthquake zone in Canada (page 120):

- About 200 km off the west coast of Vancouver Island, the Juan de Fuca and Pacific Plates are spreading apart.
- Closer to the coast, the Juan de Fuca Plate moves beneath the North American Plate along the Cascadia subduction zone.
- Off the west coast of the Queen Charlotte Islands, the Pacific Plate is sliding past the North American Plate.

Some significant earthquakes felt in Vancouver

Date	Location	Magnitude	Comment
January 26, 1700	West of Vancouver Island	8+	Great earthquake; native villages destroyed
December 15, 1872	North-central Washington	7.4	Felt strongly in the Lower Mainland
January 11, 1909	San Juan Islands	6	Deep, felt strongly in the Lower Mainland
December 6, 1918	Vancouver Island	7	Damage on west coast of Vancouver Island
January 24, 1920	San Juan Islands	5.5	Deep, felt strongly in the Lower Mainland
June 23, 1946	Vancouver Island	7.3	Much damage on central Vancouver Island
April 13, 1949	Puget Lowland	7	Deep, much damage in Seattle and Tacoma
April 29, 1965	Puget Lowland	6.5	Deep, much damage in Seattle
November 30, 1975	Strait of Georgia	4.9	Shallow, many aftershocks
May 16, 1976	Southern Gulf Islands	5.4	Deep
April 14, 1990	Fraser Lowland	4.9	Shallow, many aftershocks
February 28, 2001	Puget Lowland	6.8	Deep, much damage in Seattle and Olympia

Earthquakes in British Columbia

The vast majority of earthquakes in B.C. occur beneath the Pacific Ocean floor (page 120). Although some of the quakes are large, they are far from major population centres and pose little threat to people and property in the region. Other earthquakes, however, do occur near our cities and towns — beneath Vancouver Island, the Strait of Georgia, the Lower Mainland, and Puget Sound. Seismologists infer that these potentially damaging earthquakes are of three types:

- *Earthquakes on faults within the North American Plate.* **Crustal earthquakes** commonly occur at depths of 20 km or less on faults that cut the North American Plate. There is, on average, one such quake in southwestern B.C. every day. Most of these earthquakes are too small to be felt, but a damaging one occurs somewhere in the region about once every ten years. The largest in this century was a magnitude 7.3 earthquake in 1946, centred on central Vancouver Island.

- *Earthquakes within the Juan de Fuca Plate.* **Subcrustal** (or inslab) **earthquakes** occur as deep as 80 km within the subducting Juan de Fuca Plate. There have been several destructive earthquakes of this type beneath Puget Sound in the historical period, most recently near Olympia, Washington, in February 2001 (page 114).

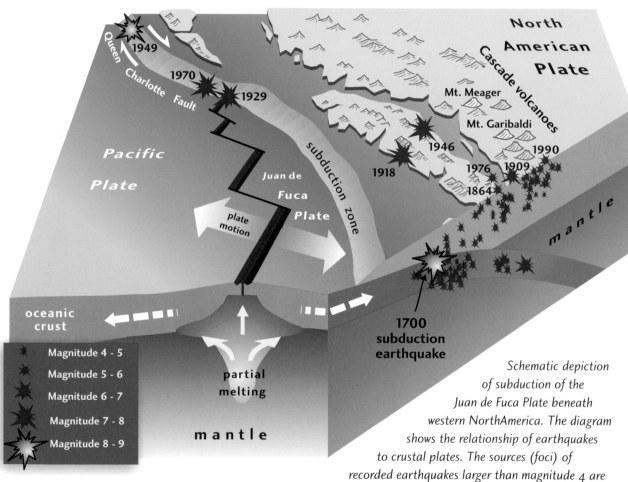

Schematic depiction of subduction of the Juan de Fuca Plate beneath western NorthAmerica. The diagram shows the relationship of earthquakes to crustal plates. The sources (foci) of recorded earthquakes larger than magnitude 4 are shown in the cross-section (only quakes close to the line of the section are depicted in the cross-section). The epicentres, or points on the Earth directly above the quake foci, are shown on the surface of the diagram. Only moderate to large earthquakes (magnitude 5 or greater) are depicted on the surface.

• *Earthquakes at the boundary between the North America and Juan de Fuca Plates.* Scientists have recently found evidence for a third, rare type of earthquake that could damage Vancouver and other cities on the south coast — the **subduction earthquake** or so-called "mega-quake." Subduction earthquakes occur along the large fault that separates the converging Juan de Fuca and North American Plates. Rupture takes place over a vast area of the fault plane and consequently produces extremely large (magnitude 8 or 9) earthquakes. There has not been such an earthquake since the first European exploration of the region, but there is abundant geological evidence for one in AD 1700 and for many others in the past few thousand years. The geological evidence is supported by geophysical measurements that show stress building up towards the next cataclysm. The data reveal that Vancouver Island is bulging slightly upwards and is contracting above the downgoing Juan de Fuca Plate, as would be expected if the two plates are stuck and not moving smoothly past each other. In a subduction earthquake, the accumulated stress is released suddenly and the plate "snaps" back, causing subsidence along the west coast of Vancouver Island.

Jan. 1700 ➤
AD 700 ➤
AD 300 ➤
600 BC ➤
1000 BC ➤

Evidence for five large earthquakes can be seen in this tidal channel at Willapa Bay, Washington. Fossil tidal marsh soils (dark horizons, indicated by arrows) represent old marsh surfaces that subsided about 1 to 2 m during earthquakes. After the earthquakes, the marsh surfaces were buried by tidal mud (light-coloured layers). The dates on the left side of the photograph are the approximate ages of the earthquakes. The uppermost buried soil records a great subduction earthquake in January 1700. During the next subduction earthquake, the modern marsh (top surface) will subside and become covered by sea water. Over time, tidal mud will bury the marsh, providing the substrate for a new marsh. The underlying sediments will record this story.

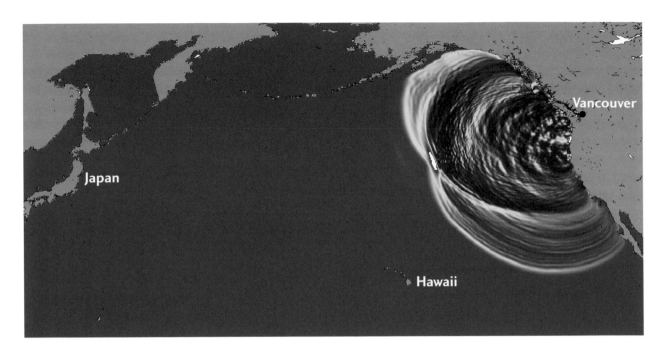

Shallow and deep — two big earthquakes in the Pacific Northwest

On the morning of June 23, 1946, a magnitude 7.3 earthquake severely shook southwestern B.C. The earthquake occurred at a depth of 30 km, with an epicentre northwest of Courtenay on Vancouver Island. Although the quake was widely felt throughout B.C., damage was limited because the area near the epicentre where the most violent shaking occurred was sparsely populated. In towns nearest the epicentre, including Cumberland and Courtenay, chimneys fell, windows broke, and walls cracked. Roads and water and utility lines were also damaged. The earthquake triggered hundreds of small landslides. Were a comparable earthquake to occur today in the vicinity of Vancouver or Victoria, damage would likely be in the billions of dollars.

A large earthquake struck the southern Puget Lowland on February 28, 2001. The quake had a magnitude of 6.8 and was centred 52 km below the Earth's surface in the vicinity of the Nisqually River delta in southern Puget Sound. Its epicentre was 17 km northeast of

Computer simulation of the tsunami generated by the great AD 1700 earthquake. The earthquake ruptured the entire 1000-km length of the boundary between the North American and Juan de Fuca Plates off the coasts of Vancouver Island, Washington, Oregon, and northern California. This simulation shows the progress of the tsunami across the Pacific Ocean three hours after the earthquake and six hours before it hit Japan. The rapidly advancing tsunami is highest in the red area. The tsunami has been precisely dated from Japanese written records. The earthquake that caused it occurred at about 9 PM Pacific Standard Time on January 26, 1700.

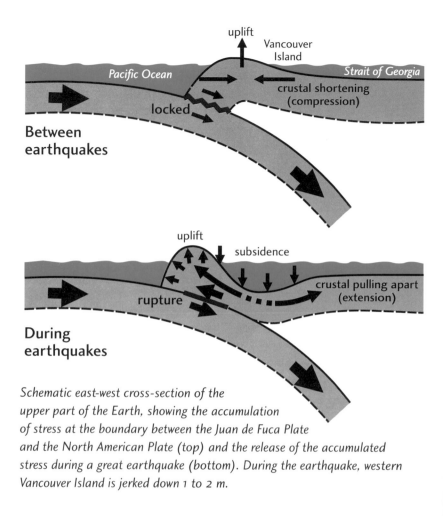

Between earthquakes

uplift

Vancouver Island

Pacific Ocean

Strait of Georgia

crustal shortening (compression)

locked

During earthquakes

uplift

subsidence

rupture

crustal pulling apart (extension)

Schematic east-west cross-section of the upper part of the Earth, showing the accumulation of stress at the boundary between the Juan de Fuca Plate and the North American Plate (top) and the release of the accumulated stress during a great earthquake (bottom). During the earthquake, western Vancouver Island is jerked down 1 to 2 m.

A fourth type of earthquake occurs off the coast of B.C., but at distances too large to cause damage to Vancouver, Victoria, or other major cities and towns in the south. These quakes occur along faults separating plates that move horizontally past one another (page 115). Geologists call these faults **strike-slip faults**. Strike-slip faults that form the boundary between two plates are termed **transform faults**. Northwest of Vancouver Island, the North American and Pacific Plates are sliding past one another along the Queen Charlotte Fault (pages 120 and 134), which is continuous

Olympia and 57 km southwest of Seattle. It caused about $2 billion in damage to roads, bridges, and buildings, destroyed 110 houses, seriously damaged another 126, and cut off power to 17,000 households. This earthquake was only slightly smaller than the quake that devastated Kobe, Japan in January 1995. The Kobe quake claimed 5500 lives and caused over $200 billion in damage and other economic losses.

Why did the two earthquakes, which both occurred in heavily populated areas, have such dramatically different effects? One reason may be that buildings and other engineered structures in the Pacific Northwest have been designed and constructed to withstand the shaking of magnitude 6 and 7 earthquakes. A more important reason, however, is that the Nisqually earthquake occurred sufficiently deep in the Earth that the intensity of the seismic waves had diminished before reaching the places where most people live — Olympia, Tacoma, and Seattle. In contrast, the Kobe earthquake was a shallow crustal event; its seismic waves travelled only a short distance from the source and struck Kobe with full force.

The great AD 1700 earthquake

It is called a "ghost forest," the stand of white skeletons of dead cedars in a tidal marsh at the shores of Willapa Bay on the southwest coast of Washington. Someone seeing this ghostly forest might ask "How could trees have grown in a tidal marsh and what killed them?" The answers to these questions were provided in the late 1980s by geologist Brian Atwater of the U.S. Geological Survey. Dr. Atwater showed that the trees were not rooted in the modern marsh, but rather in an old marsh buried by 1 to 2 m of tidal mud. He mapped out the buried surface, which he found to be marked by a thin layer of peat, and showed that it extends throughout the modern marsh. Dr. Atwater and his American and Canadian co-workers found the same peat layer in over 20 tidal marshes extending from central Vancouver Island to northern California. Carbon dating showed that the peat layer at all of the sites is about 300 years old, although exact age equivalence could not be proved because the dating technique is too imprecise.

Dr. Atwater proposed that the peat layer was an old marsh that subsided 1 to 2 m during a great earthquake about 300 years ago. Cedars growing at the edge of the marsh also subsided and were killed by brackish tidal waters that inundated their roots. The presence of the same soil over a 1000 km length of the Pacific coast supports the argument that the earthquake was very large. Using earthquakes in Chile in 1960 and Alaska in 1964 as analogs, scientists suggested that it may have had a moment magnitude of 9.

In the 1990s, scientists asked themselves "How can we date this event more precisely?" American and Japanese researchers independently tried two approaches. The Americans looked at the pattern of annual rings in trees killed or damaged by the earthquake and matched them to ring patterns of living trees many hundreds of years old that had not been affected by the quake. To use a metaphor, the procedure is a little like matching bar codes on retail products. The researchers were able to show that the outermost ring in trees killed by the earthquake dated to 1699. The Japanese researchers were even more clever. They read in one of Dr. Atwater's papers that he had found an anomalous layer of sand resting on the buried peat

with the Fairweather Fault in Alaska. In Canada, the fault is entirely offshore; its nearest landfall is the Queen Charlotte Islands. Numerous large earthquakes have occurred on the Queen Charlotte Fault, including Canada's largest historical quake, a magnitude 8.1 event in 1949.

The Queen Charlotte Fault is the same type of fault as its better-known cousin, the San Andreas Fault in California. There is, however, one very important difference — the San Andreas Fault extends through California, passing, for example, through the suburbs of San Francisco. More than 20 million people live and work within 100 km of this fault. The Queen Charlotte Fault, on the other hand, lies offshore, far from B.C.'s cities. For this, we can count our blessings!

Quake damage

Earthquake effects include strong ground shaking, ground failure and cracking, and tsunamis. Earthquakes are most destructive when centred near cities, but the damage can differ considerably from site to site because of local ground conditions and methods of building construction. For example, shaking on loose, modern sediment is generally greater than shaking on rock (**seismic amplification**). In some situations, a resonance can occur in deep sediments, markedly amplifying shaking at particular frequencies. This resonance can

layer at Willapa Bay. Dr. Atwater suggested that the sand had been deposited by a tsunami triggered by the AD 1700 earthquake. The Japanese are familiar with tsunamis and they hypothesized that a large tsunami generated near the west coast of North America would cross the Pacific Ocean and strike Japan (page 122). The researchers searched Japanese written records and discovered that a mysterious, large tsunami, which had not been accompanied by a local earthquake, occurred on January 27, 1700. The tsunami struck a 1000 km length of the eastern coast of Honshu with waves up to several metres high, causing considerable damage in some towns. The researchers concluded that the tsunami was triggered by the same earthquake that had caused the land to drop on the west coast of North America and killed the cedars shortly after 1699. They did even better — they back-calculated the travel time of the tsunami to Japan (some nine hours), factored in the change in time at the International Date Line, and suggested that the earthquake occurred at about 9 PM Pacific Standard Time on January 26, 1700! This timing is compatible with First Nations oral accounts of an earthquake and tsunami that devastated communities on the Pacific Northwest coast on a cold winter night prior to European contact.

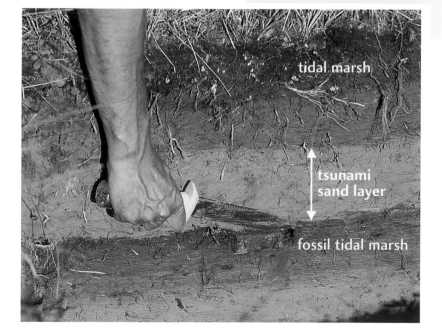

tidal marsh

tsunami sand layer

fossil tidal marsh

Bed of sand left by a tsunami in January 1700. The sand is exposed in a shallow dug pit at a tidal marsh near Tofino. It was deposited on a marsh that subsided during the earthquake and was later covered by tidal mud.

be disastrous to structures that are prone to damage at these amplified frequencies. Areas of intensified shaking may experience greater damage than nearby areas in which more consolidated Ice Age sediments or bedrock form the ground surface.

Earthquake shaking can also transform some loose, water-saturated silt and sand into fluid material (**liquefaction**). Liquefied silt and sand may move upward along cracks in the ground and erupt onto the surface to form **sand volcanoes**. At depth, the liquefied sediments lose their strength and cease to support buildings and other structures built on them. The buildings may sink or lean, or their foundations may crack due to irregular settling or horizontal movement. Materials most likely to liquefy include landfill and poorly compacted sand and silt located along the coast and beneath deltas and river plains. An area of particular concern is the Fraser River delta. A thick layer of liquefiable wet sand lies beneath a capping layer of silt and peat throughout much of the delta plain. Geologists have found widespread evidence on the delta of liquefaction triggered by a large prehistoric earthquake, and engineering tests indicate that the buried sand unit will liquefy if strongly shaken during a future earthquake.

The severity of ground motion during an earthquake depends on the thickness and physical properties of geological materials. Tall buildings located on thick, unconsolidated sediments will be more strongly shaken than those lying directly on bedrock because low-frequency, long-period seismic waves are amplified as they pass through the thick sediment pile. Tall buildings resonate with these low-frequency seismic waves, whereas small structures do not. In contrast, small buildings located on 10 to 20 m of unconsolidated sediments may be more strongly shaken than those on bedrock because high-frequency, short-period seismic waves are commonly amplified in these thinner sediments. Low buildings resonate with the high-frequency waves, whereas tall structures do not.

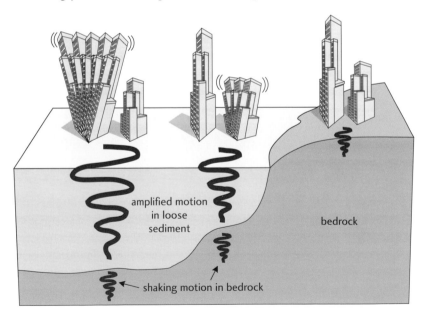

amplified motion in loose sediment

bedrock

shaking motion in bedrock

An earthquake is accompanied by rapid movement along one or more faults. The ground ruptures when this movement extends to the surface. Ground rupture can be disastrous to bridges, utility and transmission lines, dams, and other structures that cross the fault. Fault rupture zones may be very narrow (1 m or less), or the disturbed zone may be tens or hundreds of metres wide.

Liquefaction of water-saturated silt or sand during an earthquake may cause the ground to fracture and subside, damaging or destroying buildings and other human works. A layer of liquefiable sand underlies much of the Fraser delta. Curved arrows show the direction of flow of liquefied sand.

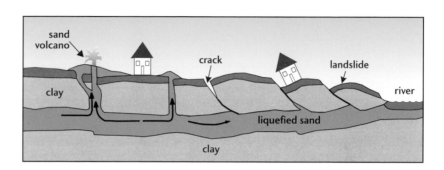

Liquefaction susceptibility map of the Fraser Lowland. Damage due to liquefaction during an earthquake is most likely in the areas coloured red, which are underlain by water-saturated, loose, silty and sandy sediments. The False Creek area is at risk because it is landfill.

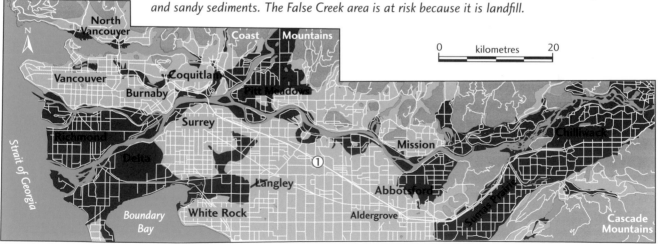

Risk of liquefaction during an earthquake

Moderate to high (modern lowland sediments and landfill)

Low (Ice Age upland sediments)

None (bedrock)

Emergency news flash!

Word has just been received that a magnitude 7.3 earthquake struck Vancouver at 10:00 AM this morning. Several hundred people in Richmond, Vancouver, North Vancouver, and Surrey have been killed by falling debris and downed powerlines, and hospitals are being flooded with the injured. Particularly hard hit are Richmond and the Gastown district in Vancouver, where some old masonry buildings have collapsed. Gas, electricity, and sewage are out over much of the Lower Mainland, and emergency crews are unable to respond because of congested roads and collapsed bridges.

The earthquake was centred 10 km beneath the floor of the Strait of Georgia 20 km west of Richmond. Damage reports are sketchy, but the Vancouver International Airport has been shut down indefinitely owing to cracking and settling of the runways, and the SkyTrain is out of service. All berths at the Tsawwassen Ferry Terminal, as well as part of the Deltaport, have slipped into the sea. The Lions Gate and Iron Workers Memorial bridges are closed due to collapse of the approaches. Numerous landslides have blocked Highway 1 and the Sea-to-Sky Highway, and the Deas Island Tunnel is flooded, effectively isolating the Lower Mainland. Fires are reported to be burning in several parts of Vancouver, Surrey, and Richmond. The sea dyke west of Richmond has failed, raising the spectre of flooding of parts of that city during the next high tide. The earthquake triggered a tsunami in the Strait of Georgia that has damaged marinas along the entire eastern Vancouver Island coastline and drowned five people at Qualicum Beach. The Prime Minister has declared the Lower Mainland a disaster area and has ordered armed forces personnel to be flown into Abbotsford to maintain order. Aftershocks, some up to magnitude 6, are causing additional damage and impeding relief efforts. Total confirmed damage is $10 billion, but is rapidly climbing and could reach $50 billion.

Following the disastrous San Fernando earthquake in 1971, the State of California passed legislation requiring that a geological site investigation be completed in "special study zones," encompassing the rupture zones of potentially and recently active faults. The purpose of the legislation is to determine whether an active fault passes through a proposed building site before a building permit is issued. The special study zones and the faults on which they based are shown on special maps. Such an approach has not been used in southwestern B.C. because faults potentially capable of strong earthquakes have not been identified in this region, at least not on land, and because there have been no instances of ground rupture during historical earthquakes. Most strong earthquakes in southwestern B.C. may be too deep to produce surface rupture.

Landslides are one of the major causes of earthquake damage and occur most commonly on steep slopes underlain by loose sediments or unstable rock. Landslides are a particularly significant seismic hazard in southwestern B.C. because steep slopes are common in the region.

Highways, rail lines, and energy transmission lines pass through valleys and canyons bordered by steep, potentially unstable slopes, and can be blocked or severed by landslides. A large earthquake might cause numerous blockages over a large area, disrupting economic activity and restricting access to the affected region. In this context, even small landslides, which are the vast majority of those triggered by earthquakes, can be severely disruptive.

Critical transportation corridors leading to the Lower Mainland that are at risk from earthquake-triggered landslides. Also shown are bridges and tunnels that could be damaged by ground shaking during a large earthquake.

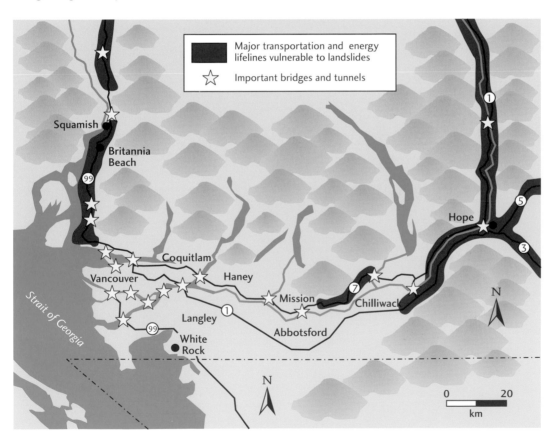

Sequential photographs of a tsunami moving ashore on the island of Oahu hours after a large earthquake off the Aleutian Islands in 1957.

Tsunamis

Some earthquakes trigger destructive sea waves, termed tsunamis. Large tsunamis can surge 1 to 2 km inland and reach heights of 30 m or more above sea level, causing loss of life and severe property damage. Earthquakes in the North Pacific are the main sources of tsunamis on our coast. The threat is greatest on western Vancouver Island; the Strait of Georgia is protected from the open ocean, thus the tsunami hazard is low at Vancouver.

Computer simulations have been made of tsunamis triggered by large earthquakes beneath the North Pacific Ocean (page 122). The simulations provide rough estimates of the size of waves that can be expected at different places on the B.C. coast. For example, for a simulated magnitude 8.5 earthquake at the northern end of the Cascadia subduction zone off Vancouver Island, maximum wave amplitudes are approximately 5 m. Waves amplify as they move up some inlets, achieving heights of 10 to 15 m at some sites. Much energy is lost as the tsunami passes through narrow passages connecting Juan de Fuca Strait, and tsunami amplitudes are reduced to a metre or less by the time the waves reach Vancouver.

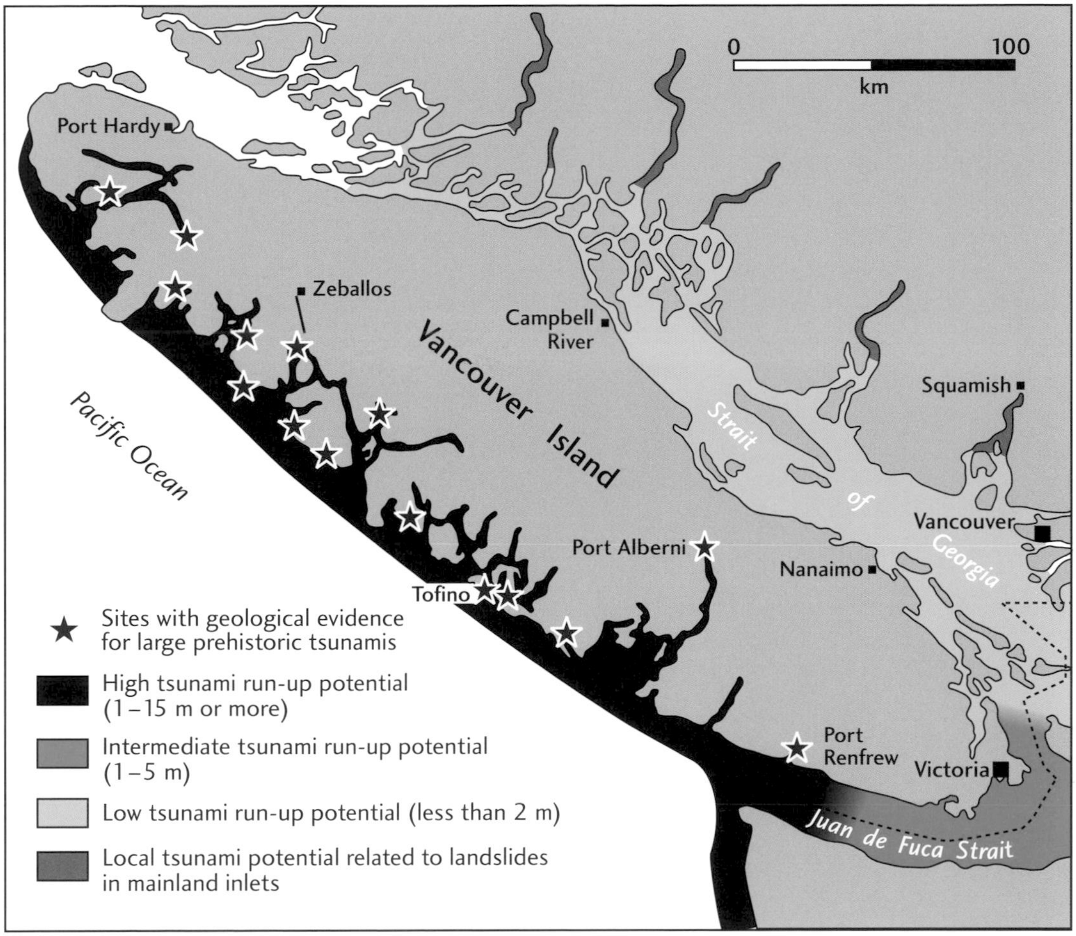
0
100
km
Port Hardy
Zeballos
Campbell River
Vancouver Island
Pacific Ocean
Squamish
Strait of Georgia
Vancouver
Port Alberni
Nanaimo
Tofino
Sites with geological evidence for large prehistoric tsunamis
High tsunami run-up potential (1 – 15 m or more)
Intermediate tsunami run-up potential (1 – 5 m)
Low tsunami run-up potential (less than 2 m)
Local tsunami potential related to landslides in mainland inlets
Port Renfrew
Victoria
Juan de Fuca Strait

Tsunami hazard zones in the coastal areas of southwestern B.C. Parts of western Vancouver Island are at greatest risk from tsunamis.

Predicting earthquakes

Occasionally, the news media report a story about someone who claims to have the power to predict when and where the next disastrous earthquake will strike. These attention seekers are charlatans who capitalize on public gullibility. When their predictions fail, these individuals slide back into anonymity, but amazingly they often surface again with new predictions.

The 1964 tsunami at Port Alberni

The Alaska earthquake of March 27, 1964, had a moment magnitude of about 9, the second largest on Earth in the twentieth century. It triggered tsunamis that killed 130 people, some as far away as California. The main tsunami swept southward across the Pacific Ocean at a velocity of over 800 km per hour (about 220 m per second), reaching Port Alberni in six and one-half hours and Antarctica in only sixteen hours. Waves were amplified as they moved up Alberni Inlet and, consequently, were two and one-half times larger at Port Alberni than at Tofino and Ucluelet on the open coast.

Three main waves struck Port Alberni between 12:20 AM and 3:30 AM on March 28. Most people in the town were asleep when the first wave rolled in. The sea surged up Somass River at a velocity of about 50 km per hour and spilled onto the land, inundating whole neighbourhoods in chest-deep water. This first wave peaked at 3.7 m above mean sea level and knocked out the Port Alberni tide gauge.

The second and most destructive wave swept into town less than two hours later, at 2 AM. The lights of the waterfront mills went out as the water smashed through the facilities. The ground floor of the

In truth, we are decades away from being able to predict earthquakes. Worse, earthquake prediction, to be useful, will have to be almost 100 percent accurate. False warnings followed by precautionary measures such as evacuations and business closings are unacceptable in a democratic society. Research on precursor signals of earthquakes, however, continues in many countries. The Chinese, for example, record unusual animal behaviour because they have noted that many animals, including dogs, rats, and

THE FAR SIDE® BY GARY LARSON

The mysterious intuition of some animals

birds, sense some change in the Earth hours or days before some earthquakes. Japanese researchers have observed that radon gas concentrations in some wells increase before large earthquakes. And scientists in many countries have installed and monitored instruments capable of measuring displacements as small as millimetres on opposite sides of active faults to determine if quakes are preceded by unusual movements.

Although we cannot predict earthquakes, we do know what parts of the world are at risk from earthquakes, how likely earthquakes are to occur in these areas, and what the damaging effects of these earthquakes will be. The geological record tells us, for example, that subduction earthquakes of magnitude 8 or 9 occur, on average, once every 500 years along our coast. On the basis of the instrumented record of earthquakes, we can expect an average of one earthquake of magnitude 7 somewhere in the Pacific Northwest about every 30 years.

Role of geology in minimizing risk

A large earthquake can cost billions of dollars and claim thousands of lives. Although we cannot prevent earthquakes, the damage they cause can be reduced through geological and geophysical studies, proper design of buildings, planning, and public education. For example, areas of possible severe ground shaking, liquefaction, or landslides can be identified through geological mapping. Mapping can aid in urban planning and in developing emergency procedures for earthquakes. Continuous monitoring of seismic activity by the Geological Survey of Canada improves our understanding of earthquake hazards in our region.

Barclay Hotel, 1 km inland, was splintered by the surging water; guests had to be plucked from an upper floor by police in boats. Logs and debris crashed into buildings, and houses were swept off their foundations and hurtled inland. As the water subsided, some buildings were dragged seaward; two houses drifted into Alberni Inlet and were never seen again. The second wave left a mark on the tide gauge at 4.3 m above mean sea level.

The third wave, which arrived at about 3:30 AM, was the largest of all, but because the tide had fallen it crested at 3.9 m and did little further damage. Other waves oscillated in Alberni Inlet with decreasing strength for another two days.

Two hundred and sixty homes in Port Alberni were damaged by this tsunami, and the total economic losses here and elsewhere on Vancouver Island were estimated at about $10 million.

As destructive as it was, this event pales in comparison to some other historical tsunamis in the Pacific Ocean. The eruption of Krakatoa volcano in 1883, for example, triggered a tsunami that killed about 37,000 people in Indonesia, and in 1896 earthquake-triggered waves up to 35 m high struck the east coast of Japan, smashing more than 100,000 houses and drowning 26,000 people.

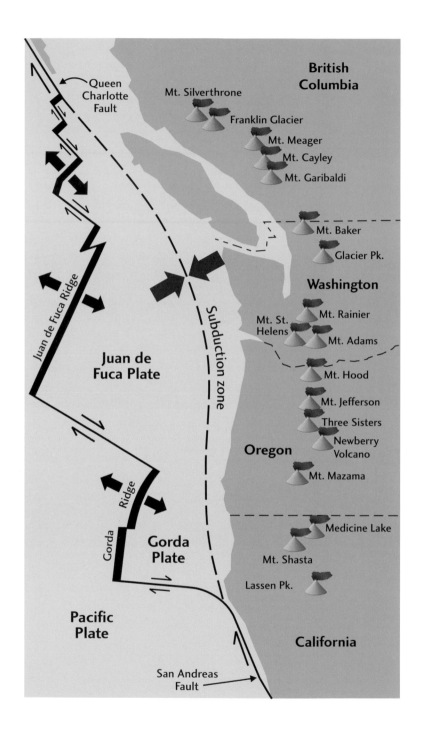

Active, dormant, and recently extinct volcanoes of the Cascade volcanic chain. Cascade volcanoes are products of subduction of oceanic crust beneath western North America.

The two red arrows show the direction of convergence of the Juan de Fuca Plate with the North American Plate. The Juan de Fuca Plate is moving down to the east under North America. The volcanoes lie above the zone where the descending plate heats to a critical temperature, causing overlying crustal rocks to melt.

The heavy black arrows indicate directions where plates are pulling apart at oceanic ridges. The thin black arrows indicate direction of relative motion where plates are sliding past one another.

In the shadow of volcanoes — volcanoes in the Pacific Northwest

ABOUT 150 VOLCANOES in B.C. and Yukon are young in a geological sense, having erupted one or more times during the past two million years. Although none of these volcanoes erupted in the twentieth century, a lava flow in the Unuk River valley in northwestern B.C. is less than 200 years old, and a flow in the Nass River valley north of Terrace is only a few hundred years old. The only active volcano close to Vancouver is Mount Baker, an icon of our local landscape.

The Cascade volcanic chain

A chain of active and recently active volcanoes extends along the Cascade Range and the southern Coast Mountains from northern California to southwestern B.C. The chain includes such well known volcanoes as Mounts Lassen, Shasta, Hood, St. Helens, Rainier, and Baker. Mounts Garibaldi, Cayley, and Meager are the most important Canadian members of the group. Many of the volcanoes of the Cascade chain are **stratovolcanoes** that have formed through repeated explosive eruptions of silica- and volatile-rich magmas. Explosive eruptions from these volcanoes in the future could endanger people and property hundreds of kilometres away.

What is the explanation for this chain of volcanoes? In a word, subduction. The Juan de Fuca Plate gets hotter as it slowly moves downward beneath the crust of North America. At a depth of 80 to 100 km, the subducting plate reaches a temperature of about 800° C, hot enough to cause mineral transformations that release

Volcanoes — one size does not fit all

Active volcanoes come in different shapes and sizes, reflecting the types of eruptions that form them. Beautiful conical volcanoes, such as Mount Fuji and Mount Baker, have been created by numerous explosive eruptions similar to the May 1980 eruption of Mount St. Helens in Washington. The magmas responsible for these stratovolcanoes are rich in silicon, sodium, aluminum, and water, and they produce volcanic rocks with names such as **andesite**, **dacite**, and **rhyolite**. These magmas are more viscous (stiffer) than magmas with lower silicon content and higher magnesium and iron contents that form basalt. Their high viscosity and fluid content make them highly explosive when they rise up into the throat or neck of a volcano. Take the pressure off and they literally explode — gases bubble out of solution, much as carbon dioxide effervesces when the lid is removed from a soft-drink bottle. Rapidly

Mount Baker, an active volcano, looms above the Vancouver skyline. It last erupted in the 1800s. A major eruption would melt the glaciers on Mount Baker, causing floods and lahars in the valleys that drain the mountain.

superheated fluids into the lower part of the overlying North America crust. These superheated fluids melt some of the crustal rocks, producing magma that moves upward along deep fractures. The magma may reach the Earth's

surface where it is erupted as **lava** or hot fragmental (**pyroclastic**) ejecta. Many of the volcanoes of the Cascade chain have a long history of eruptions, which indicates that the areas of melting deep in the crust, as well as the volcano's "plumbing system," have not changed much over hundreds of thousands, or even millions, of years.

Mount Baker, the sleeping giant

Glacier-cloaked Mount Baker reaches an elevation of 3285 m in the northernmost Cascade Range of Washington. The mountain is a great pile of ash, lava, and debris left by numerous eruptions over the last 30,000 years. The relatively undissected conical shape of Mount Baker reflects its status as an active volcano and distinguishes it from more eroded, inactive volcanoes like Mount Meager and Mount Garibaldi. Indeed, conical-shaped volcanoes around the world, such as Fuji in Japan and Popocatépetl in Mexico, are without exception active. If they were not, erosion would long ago have eroded the beautiful cones into more irregular shapes. Mount Baker most recently erupted in the mid-1800s. Venting of gases and hot fluids from the summit crater in the late 1970s provides a timely reminder that the volcano is still active.

Hazard assessment has been done at Mount Baker and other active volcanoes of the Cascade chain to provide long-term planning in areas likely to be affected by a future eruption. The hazards of greatest concern are landslides, volcanic debris flows (**lahars**), and filling of river valleys with sediment (**aggradation**).

effervescing gases expand and fragment the magma, propelling it out of the volcano like a bullet from a gun. Large plumes of **tephra** rise in billowing clouds kilometres into the atmosphere, blanketing the ground with **ash** far from the volcano. Red-hot blocks fall on the flanks of the volcano and trigger incandescent flows that surge down slope at high speeds, vaporizing all living things in their path. You get the idea — these eruptions are very nasty customers! Stratovolcanoes are common at subduction zones where continental crust is melted above a hot, subducting oceanic plate.

A very different type of volcano is produced by basaltic magmas, which have relatively low silicon, high magnesium, and high iron contents. Basaltic magmas contain less water and are hotter than dacitic and andesitic magmas. As a consequence, they are less viscous and less explosive; they typically erupt lavas that flow relatively quickly down the volcano's flanks. Successive eruptions over periods of hundreds of thousands to millions of years construct broad, rounded volcanoes, termed **shield volcanoes**. Classic examples of shield volcanoes are Mauna Loa and Mauna Kea on Hawaii. Shield volcanoes do not occur at subduction zones. Instead, they are found where basaltic magma wells to the surface from the mantle without significant contamination by continental crust, for example in Iceland and on the Hawaiian Islands. Good examples of shield volcanoes in B.C. include Mount Edziza in the Stikine River area and the Ilgachuz Range in the Chilcotin. Such eruptions pose much less of a risk to us than explosive eruptions of stratovolcanoes, unless of course we have the very bad luck of being in the path of a lava flow.

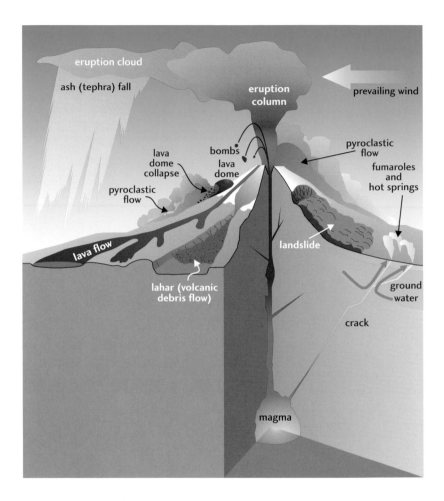

The following labels appear in the diagram:

eruption cloud

ash (tephra) fall

bombs

eruption column

prevailing wind

lava dome collapse

lava dome

pyroclastic flow

pyroclastic flow

fumaroles and hot springs

lava flow

landslide

lahar (volcanic debris flow)

ground water

crack

magma

Hazards associated with explosive volcanoes such as Mount Baker and Mount Meager.

The rocks around Mount Baker's summit crater have been extensively altered to soft, slippery clay minerals by circulating hot acidic groundwater. They could fail if the volcano became inflated with magma at the onset of an eruption. Landslides could also occur on the steep flanks of the volcano at other times, without an accompanying eruption. Such landslides range in size from small movements of loose debris on the volcano's flanks to massive collapses of the entire summit or sides of the volcano. Mount Baker and other Cascade volcanoes are susceptible to landslides because they are built up partly of layers of loose volcanic rock fragments.

Chronicle of a volcano: Mount Baker's last 12,000 years

Date	Volcanic Activity
1975 to present	*Steam activity*
Mid-1800s	*Small explosive eruptions create fiery displays visible at night in Victoria*
Last several centuries	*Two lahars travel 9 and 11 km*
500 to 600 years ago	*Lahar travels 14 km*
6800 years ago	*Large lahar reaches Puget Sound*
7600 to 12,000 years ago	*Eruption of tephra and lava, lahar*
9600 years ago	*Eruption of tephra, pyroclastic flows, lava flows, lahars*
12,000 years ago	*Eruption of lava*

During an eruption of Mount Baker, hot volcanic debris would mix with water melted from snow and ice on the summit and flanks of the mountain to form large debris flows, or lahars. These flows of mud, rock, and water rush down river valleys at speeds of up to 70 km per hour and can travel more than 100 km. Some lahars contain so much rock debris that they look like fast-moving rivers of wet cement. Close to their source, lahars are powerful enough to rip up and transport trees and huge boulders. Farther downstream, they entomb everything in their path in muddy debris. In the worst case scenario, a large lahar might reach the Bellingham area, as apparently happened about 7000 years ago. Lahars might also enter Baker Lake, a reservoir near the base of Mount Baker, and displace enough water to overtop the dam or to cause it to fail. Failure of the dam would be catastrophic for downstream communities along the Skagit River in Washington. A lahar or series of lahars moving down the Nooksack River on the northern side of Mount Baker could deposit enough material to force water or debris into the eastern Fraser Lowland at Sumas, or even divert the entire river into the Fraser Lowland.

Large landslides or lahars would increase sediment loads to rivers such as the Skagit and Nooksack and cause them to build up their beds. Settlements and other infrastructure on the valley floors might be buried or flooded as the level of the river beds rose.

Other volcanic hazards

Explosive ash eruptions

A large volcanic debris flow (lahar) moving down the Nooksack River valley from Mount Baker might spill into B.C. at Sumas.

An explosive eruption blasts solid and molten rock fragments and volcanic gases into the air with tremendous force. The largest rock fragments, referred to as **bombs**, generally fall to the ground within 3 km of the vent. Small particles (ash) rise high into the air, forming a huge, billowing **eruption cloud**. An eruption cloud can pose a serious hazard to aviation. During the past fifteen years, about eighty commercial jets have been damaged by inadvertently flying

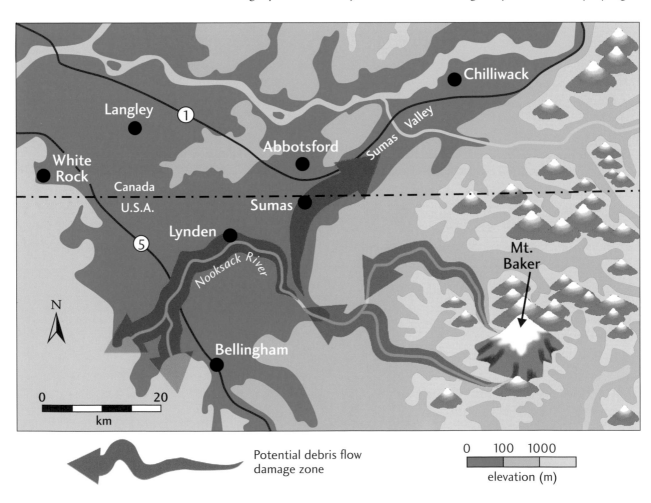

Potential debris flow damage zone

0   100   1000
elevation (m)

into clouds of ash, and several have nearly crashed because of engine failure. Large eruption clouds can extend hundreds of kilometres downwind, depositing ash over enormous areas. Ash from the May 18, 1980, eruption of Mount St. Helens fell over an area of more than 50,000 square kilometres in the western United States; small amounts of ash even fell in southernmost B.C. Heavy ashfall can collapse buildings, and even minor amounts of ash can cause respiratory problems and damage crops, electronics, and machinery.

Is Vancouver at risk from an ash fall from an eruption of Mount Baker? The fallout area is determined by the direction of the wind at the time of the eruption. In the Vancouver area, winds are dominantly from the west to the east. If winds were blowing in this direction during an eruption, which is the most probable scenario, much of the ash and dust would likely be carried into eastern Washington and perhaps south-eastern B.C. If, however, the winds were from the southeast, as happens during some winter storms, the Vancouver area could receive a significant dump of Mount Baker ash — not a very pleasant white Christmas!

In the past 12,000 years, other volcanoes in the Cascade chain have erupted more than 200 times, and several of these eruptions have deposited significant quantities of ash in southern B.C. However, Vancouver is positioned slightly west of the volcanic chain, and prevailing high-altitude winds blow toward the east. As a result,

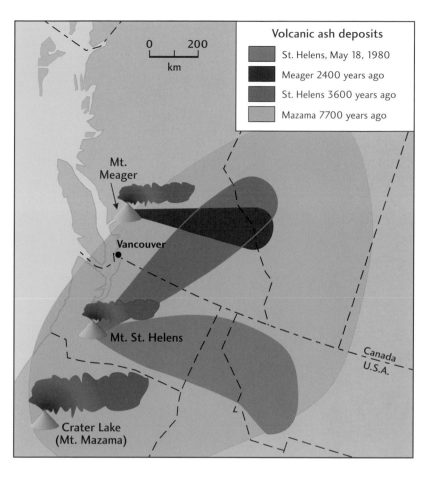

Source and distribution of some widespread volcanic ash deposits found in southwestern Canada and northwestern U.S. The pattern of each deposit reflects the size of the volcanic eruption and the direction of the prevailing wind at the time of the eruption.

the city is less likely to receive significant amounts of ash than communities in the B.C. interior. Scientists have estimated that 10 cm of ash can be expected to fall in the Fraser Lowland, on average, once every 10,000 years. More frequent, smaller falls of ash can also be expected; the most likely sources are Mount St. Helens and Mount Baker.

Dangerous gases

Volcanoes emit gases during eruptions. Ninety percent of the gas is water vapour (steam), most of which is heated groundwater. Other common volcanic gases are not so benign — carbon dioxide, carbon monoxide, sulphur dioxide, hydrogen sulphide, hydrogen, and fluorine. Sulphur dioxide can react with water droplets in the atmosphere to create acid rain, which causes corrosion and harms vegetation. Carbon dioxide is denser than air and can be trapped in areas in low-lying areas in concentrations that are deadly to people and animals. Fluorine is toxic in high concentrations and can be incorporated into volcanic ash particles that later fall to the ground. The fluorine in the particles can poison livestock and contaminate domestic water supplies.

Cataclysmic eruptions, such as the June 15, 1991, eruption of Mount Pinatubo in the Philippines, inject huge amounts of sulphur dioxide gas into the stratosphere, where it combines with water to form an aerosol of sulphuric acid. Such aerosols and volcanic dust reflect solar radiation and can lower the Earth's average surface temperature by up to 2° C for extended periods of time. Sulphuric acid aerosols also damage the ozone layer by altering chlorine and nitrogen compounds in the upper atmosphere.

Basaltic lava fountain and flow, Kilauea volcano, Hawaii. Eruptions of this type produced the lava flows at Brandywine Falls north of Squamish and the Ring Creek and Clinker Peak flows at Mount Garibaldi.

Flowing lava

Magma that pours or oozes onto the Earth's surface is called lava. The higher the silicon content of a lava, the less easily it flows. Low-silica basalt can form fast-moving flows or can spread out in broad thin sheets up to many kilometres wide. In contrast, higher-silica andesite, dacite, and rhyolite lavas are typically thick and sluggish, travelling only short distances from a vent. In this context, the Ring Creek lava flow east of Squamish is unusual because, in spite of its relatively high content of silicon, it was able to flow 17 km down a valley from its source vent.

*Computer-generated image of the Ring Creek lava flow, east of Squamish. The flow issued from Opal Cone (in red) after the end of the Ice Age and flowed about 17 km down the valley to the Mamquam River. Note the raised ridges, or **levees**, at the margins of the lava flow. The levees formed because the flow margins cooled and solidified while the main body of lava continued to flow.*

Brandywine Falls

Brandywine Falls is a short walk from the parking lot at Brandywine Falls Provincial Park, on Highway 99 about 47 km north of Squamish. The walk is well worthwhile. During times of high flow, the falls are spectacular — Brandywine Creek plunges 66 m over the lip of a series of lava flows to the valley below. Each of the several thick horizontal layers visible in the photo represents one lava flow. The lava flows are more resistant to erosion than the rocks beneath them (much like Niagara Falls), thus the falls are maintained as they slowly retreat back upvalley.

Glowing avalanches

Avalanches of incandescent ash, rock fragments, and gas can move down the flanks of a volcano during explosive eruptions or when the steep side of a growing **lava dome** collapses and breaks apart (page 138). These **pyroclastic flows** are as hot as 800° C and move at speeds of 150 to 250 km per hour. The flows tend to follow valleys, knocking down and incinerating everything in their paths. Low-density pyroclastic flows, called **pyroclastic surges**, can overtop ridges more than 100 m high. The bulk of Mount Garibaldi consists of deposits of pyroclastic flows erupted at the end of the Ice Age. These deposits are evidence of a series of violent eruptions. If a similar eruption were to occur today, Squamish would be no more.

The climactic eruption of Mount St. Helens on May 18, 1980, produced a huge pyroclastic surge, the so-called "lateral blast" that destroyed an area of 600 square kilometres. Trees 2 m in diameter were mowed down like blades of grass as far as 25 km from the volcano.

Mount Meager — still cooking!

The Mount Meager complex northwest of Pemberton is the northernmost volcano of the Cascade volcanic chain. It is a dormant volcano made up of lava flows, volcanic ash layers, and rubble. Although less than 2 million years old, Mount Meager is deeply eroded and lacks the classic cone shape of active volcanoes. The last known eruption occurred about 2400 years ago (400 BC), several hundred years before the Roman civilization reached its peak. High temperatures in rocks at shallow depths beneath the mountain indicate that magma still reaches high in the crust here and that the volcano is not dead (page 261).

The last eruption of Mount Meager was an explosive one. A great cloud of dust and ash was blown high into the atmosphere and carried east by winds across B.C. Fragments of **pumice** the size of golf balls rained out close to the volcano, covering the ground to depths of many metres in some places. Farther east, the particles decrease in size and the deposit thins markedly. Remnants of the thin ash layer produced by this 2400-year-old eruption occur in a band extending from Mount Meager across southern B.C. into westernmost Alberta (page 141).

The eruption also produced a flow of incandescent ash and lava blocks that filled the Lillooet valley at the foot of the volcano. This flow blocked the Lillooet River and impounded a large lake upstream from the dam. Shortly afterwards, the lake either overtopped and rapidly incised the lava dam, or simply caused it to collapse. A flood of water and hot volcanic debris surged down the valley devastating everything in its path. The entire Lillooet River valley, as far as Lillooet Lake, and including the Pemberton area, would be at risk from floods and lahars during a future eruption of Mount Meager.

Pylon Peak in the Mount Meager massif, a deeply eroded Quaternary volcano in the southern Coast Mountains. This volcano is the source of southwestern B.C.'s most recent volcanic eruption, which occurred 2400 years ago. Each of the conspicuous, horizontal layers visible near the summit of the peak represents a single explosive eruption. Two huge landslides, the first about 8700 years ago and the second about 4400 years ago, created the amphitheatre-like bowl below Pylon Peak, outlined by the dashed red line. See page 161 for a cut-away view of this volcano.

Mount Garibaldi — fire and ice

Mount Garibaldi, located about 65 km north of Vancouver near the town of Squamish, is a deeply dissected, dormant volcano. A viewpoint on Highway 99, 1.5 km northeast of Browning Lake (Murrin Provincial Park) and 3 km southwest of Shannon Falls, provides a particularly good view of the volcano if the weather is fine.

The volcano's long history of eruptions extends back more than 2 million years. The most recent eruption, about 12,000 years ago,

Tephra with pumice fragments up to cobble size, deposited during the last eruption of Mount Meager, 2400 years ago. The photo was taken near the pumice mine in the Lillooet River valley, close to the source of the eruption. The ash layer here is about 3 m thick. Ash from this eruption has been found as far east as Alberta.

produced the Ring Creek lava flow in the Mamquam River valley east of Squamish. Other, slightly older flows, which issued from vents on the flank of Mount Garibaldi, show evidence of having erupted when glaciers still filled nearby valleys. For example, a lava flow from Clinker Peak, a cone on the flank of Mount Price, came into contact with the decaying Cheakamus valley glacier at The Barrier (pages 148 and 149). There, the chilled front of the lava flow froze and formed a steep rock face several hundred metres high. A trail from a parking lot at the head of Rubble Creek, which can be reached by paved road from Highway 99 just south of Daisy Lake, provides magnificent views of The Barrier.

Since the Cheakamus valley glacier disappeared about 11,000 years ago, the fractured face of The Barrier has repeatedly collapsed, producing large landslides in Rubble Creek valley. The most recent of these landslide dates to the winter of 1855–1856; its bouldery deposits are crossed by the Highway 99 just south of Daisy Lake. Trees killed and buried in the deposits of the landslide have been exposed by river erosion and can be seen along the banks of the Cheakamus River for a kilometre downstream from the mouth of Rubble Creek. The dam impounding Daisy Lake is itself partly built on landslide deposits. A future landslide from The Barrier could block Highway 99, damage the Daisy Lake dam, and displace water from Daisy Lake over the dam, causing flooding in the Cheakamus River valley and the Squamish area.

In 1981, the Government of British Columbia designated the Rubble Creek area as being too hazardous for human habitation and set aside $14 million to buy out and relocate property owners in the area. This decision was made after lengthy litigation by a developer who had proposed building a subdivision on the Rubble Creek fan. Knowing that a large landslide occurred here in the mid-1800s and that more are likely in the future, would you live here?

Remnants of a forest buried by coarse gravel during the 1855–1856 Rubble Creek landslide. The view is up Rubble Creek from Cheakamus River to The Barrier (the cliff in the background). Rubble Creek and Cheakamus River have cut down through the gravel and exhumed the trees.

The Table

Clinker Peak

Garibaldi Lake

levees

lava flow

Lesser Garibaldi Lake

The Barrier

Barrier Lake

Map of the Garibaldi Lake area, showing Clinker Peak, near the summit of Mount Price, and the lava flow that issued from it about 12,000 years ago. The lava flow created the basin in which Garibaldi Lake lies, as well as The Barrier. The grey lines are contours, which are lines of equal elevation. In this case, the vertical spacing between contours is 200 m. The road up Rubble Creek to the parking lot is built on the 1855–1856 landslide. The hiking trail to Garibaldi Lake (black dashed line) was constructed well above the path of the landslide. Garibaldi, Lesser Garibaldi, and Barrier Lakes are ponded behind the flow that created The Barrier. The lava is highly fractured, and for much of the year the lakes drain from one to the other by underground channels. At times of high flow, Garibaldi Lake flows into Lesser Garibaldi Lake by an intermittent stream (dashed blue line). Barrier Lake drains underground, and Rubble Creek emerges full-blown from landslide debris at the base of The Barrier. The possibility of future collapse of The Barrier is sufficiently high that development has been prohibited in the area near the mouth of Rubble Creek.

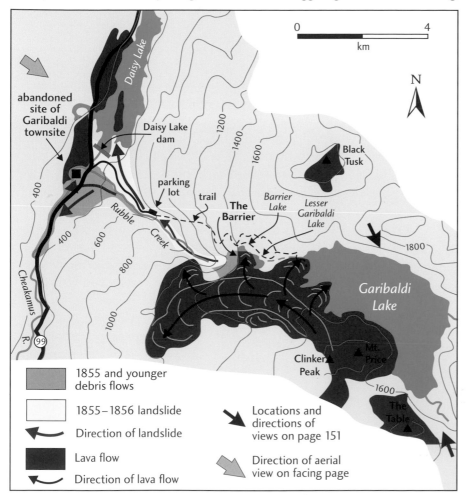

0   4
km

N

Daisy Lake

abandoned site of Garibaldi townsite

Daisy Lake dam

parking lot   trail

The Barrier

Barrier Lake   Lesser Garibaldi Lake

Black Tusk

1800

Rubble Creek

Cheakamus R.   99

Garibaldi Lake

Clinker Peak   Mt. Price

1600

The Table

Legend:

- 1855 and younger debris flows
- 1855–1856 landslide
- Direction of landslide
- Lava flow
- Direction of lava flow
- Locations and directions of views on page 151
- Direction of aerial view on facing page

Opposite: The Barrier is a steep, unstable face several hundred metres high that formed when the Clinker Peak lava flow came into contact with an Ice Age glacier in Cheakamus valley. The Barrier is the source of the 1855–1856 Rubble Creek landslide. The flat-topped Table, in the background, is a pillar of flat-lying lava flows that erupted into the base of the ice sheet that covered British Columbia at the end of the Ice Age about 15,000 years ago. Much of the scenery of Garibaldi Park is a story of "fire and ice."

Garibaldi Lake

Garibaldi Lake can be reached by a good trail from the parking lot at the head of Rubble Creek. The lake is up to 250 m deep and is dammed by a lobe of the Clinker Peak lava flow. The blue colour of the lake results from the presence of rock particles (**rock flour**) suspended in the water. The particles preferentially reflect sunlight with wavelengths of blue light. Rock flour is created by the grinding of rock at the base of glaciers and is carried into lakes by sediment-laden streams.

Part of Mount Garibaldi was constructed on a glacier. The edifice consists of block and ash deposits emplaced by hot pyroclastic flows. The western half of the volcano is missing and likely collapsed into Squamish valley when the glacier onto which it erupted melted away. The collapse exposed the edifice's blocky pyroclastic flow deposits and created the steep, rugged slope at the head of the Cheekye River. This slope is extremely unstable — recurrent landslides from the face have produced the large debris fan at the mouth of the Cheekye River. Highway 99 ascends the south flank of this fan just north of the community of Brackendale. Much of the upper part of the Cheekye fan is forested, but as Squamish and Brackendale have grown, pressure to develop the fan has increased. However, the possibility that a large debris flow could spill onto the fan caused the District of Squamish to adopt regulations that limit or restrict development in the area.

Another interesting example of an eruption of lava against an ancient glacier can be seen at the BC Rail rock quarry on the eastern side of Highway 99 near Callaghan Creek, 5 km southwest of Function Junction on the southern outskirts of Whistler. Trespassing is not allowed, but BC Rail cannot stop you from taking pictures from the fence. What you immediately notice is that the basalt in the quarry exhibits **columnar jointing** . Columns form in basalt when magma cools and crystallizes. As the magma solidifies, its volume decreases and polygonal cracks develop. Cracks in most columnar-jointed basalt flows are vertical because cooling is most rapid perpendicular to the surface over which the lava flows. The columns in parts of the BC Rail quarry, however, point every which way — up, down, and sideways. This helter-skelter pattern of

Columnar-jointed basalt exposed in the BC Rail quarry along Highway 99 about 5 km south of the southern outskirts of Whistler. Note the variable orientation of the basalt columns.

columns suggests that the lava cooled in contact with ice. From the air, the lava flow can be seen to have a sinuous form, much like an **esker**. This and the crazy-quilt pattern of columns led Professor William Mathews, one of the deans of B.C. geology, to conclude that the lava erupted beneath a glacier and flowed through a tunnel at the base of the ice.

It is intriguing that so much volcanic activity at Mount Garibaldi occurred at the end of the Ice Age and that the volcano has been inactive since then. The ice sheet that buried the volcano as recently as 15,000 years ago disappeared about the time the volcano became active. Perhaps the removal of this ice load deformed the crust enough to cause magma to move up into the volcano, initiating a series of eruptions.

Two photos of Garibaldi Lake. The photo above is a view southeast down the lake. Mount Garibaldi is the high peak on the right, with the flat-topped Table beneath it. The photo below is a view north-northwest across the lake to the sharp peak of The Black Tusk, the eroded core of a Quaternary volcano.

GOLD **URANIUM** **POTASH** **HALITE** **COAL** **CLAYS** **SILICA SAND** **SODIUM SULPHATE** **POTASSIUM SULPHATE** **SAND & GRAVEL**

MINERALS AT WORK HOME AND PLAY

SHIRT
Petro-
chemicals

HELMET
Petrochemicals
and Iron

ELBOW PADS
Copper, Zinc,
Petrochemicals

WATCH
Silicon, Copper,
Chromium, Iron,
Nickel, Silver,
Mercury, Carbon,
Zinc, Lead, Tin

GLOVES
Petrochemicals

CASSETTE/RADIO
Silicon, Copper,
Chromium, Iron,
Nickel, Silver,
Mercury, Carbon,
Zinc, Lead, Tin

**POCKET
ZIPPER**
Aluminum

SHORTS
Petro-
chemicals

KNEE PADS
Copper, Zinc,
Petrochemicals

ROLLER BLADES
Petrochemicals,
Iron, Chromium,
Nickel, Aluminum

Earth resources

I F YOU ARE indoors at this moment, take a look around you. Whether at home, in an office, or in your car, most of what you see has been extracted from the Earth. The metals in your car, appliances, computer, household furnishings, and office buildings are derived from bodies of ore that are mined. Plastics and other petrochemical products used in your carpets, cosmetics, cars, computers, most of your clothing, some furniture, packaging material, fibre-optic cables, and water and sewer lines are produced from petroleum that is pumped from the Earth. Glass is produced from quartz, a common Earth mineral. Without petroleum and natural gas, you would not be able to drive your car or take a bus. In short, without Earth materials, civilization would not exist.

Most of us live in cities and have limited awareness of our connection to the Earth. As a result, we may not appreciate the necessity of extracting Earth resources to maintain life as we know it. Mining is viewed by some as a "dirty," or "sunset" industry, something to be done in developing countries but not in Canada. To be blunt, such attitudes are uninformed and hypocritical. Mining is not a luxury; it is essential. Of course, it must be done more carefully and with much fewer adverse environmental impacts than in the past.

Ore concentrator at the Britannia Mine, now the site of the B.C. Mining Museum.

Facing page: Minerals and petroleum are required for most of what we use at work, in the home, and at play. From a poster entitled "Minerals at Work Home and Play," courtesy of Saskatchewan Industry and Resources (formerly Saskatchewan Energy and Mines).

Britannia Mine — an economic driver

The mill of the old Britannia Mine south of Squamish is a familiar sight to motorists on the Sea-to-Sky Highway. The mill processed metal-liferous rock from the Britannia **orebody** that lies within the mountain behind the mill buildings. From 1905 to 1974, about 50 million tonnes of sulphide-rich rock containing, on average, 1 percent copper, 0.6 percent zinc, and other metals such as lead, silver, gold, and cadmium, passed through this mill. At its peak, the Britannia Mine employed over 2000 people. Over its life, the mine produced metal that today would be worth $1.3 billion.

Copper is present in small amounts in almost all rocks, but at Britannia it became greatly concentrated by natural processes. The part of a copper deposit that can be mined is referred to as the orebody. The mine consisted of the infrastructure — tunnels, open pits, buildings, and waste dumps — required to extract the ore and dispose of the waste rock. Separation of copper-bearing minerals from the waste rock at Britannia was done by crushing and density flotation in the mill buildings. Pulverized waste rock from the mill was carried by slurry pipeline and discharged into Britannia Creek and the waters of Howe Sound.

The Britannia orebody comprises a series of sulphide-rich lenses that extend over 2 km from the top of Britannia Mountain to below sea level and over 4 km in an east-west direction. Part of the orebody was exposed at the surface near the top of the mountain, forming rusty, green-stained outcroppings that were discovered by a Dr. Forbes while deer hunting in the 1890s.

The origin of the Britannia Mine has been a subject of debate among geologists over the years. The ore minerals occur within a fault zone, and movements along the fault extensively deformed the orebody, obscuring many details of its origin. Geological investigation revealed that the orebody lies within rocks formed by explosive volcanic eruptions. These volcanic rocks are interlayered with mudstone that contains marine fossils and thus must have erupted on an ancient seafloor. Paleontologists have dated the fossils to the middle part of the Cretaceous Period, about 105 million years ago. Nearby granitic rocks were thought by some geologists to be the source of the copper, but they cut across, or intrude, the volcanic rocks and are therefore younger than the orebody. During the 1970s, comparisons were made between the Britannia ores and metal deposits currently forming at submarine hot springs along modern ocean ridges. The similarities are compelling and have led most geologists to conclude that the Britannia orebody formed around hot springs below a Cretaceous sea. Much has changed in the last 105 million years!

*Facing page: The complex evolution of the Britannia orebody from seafloor hot spring to the B.C. Mining Museum of today. Top left: origin of the Britannia metal deposit on the seafloor adjacent to "**black smokers**" (next page) about 105 million years ago. Top right: burial of metal deposit by lava and mud. Bottom left: compression and faulting of metal deposit during mountain building; intrusion by granitic magma about 100 million years ago. Bottom right: uplift and erosion has exposed metal deposit at the surface.*

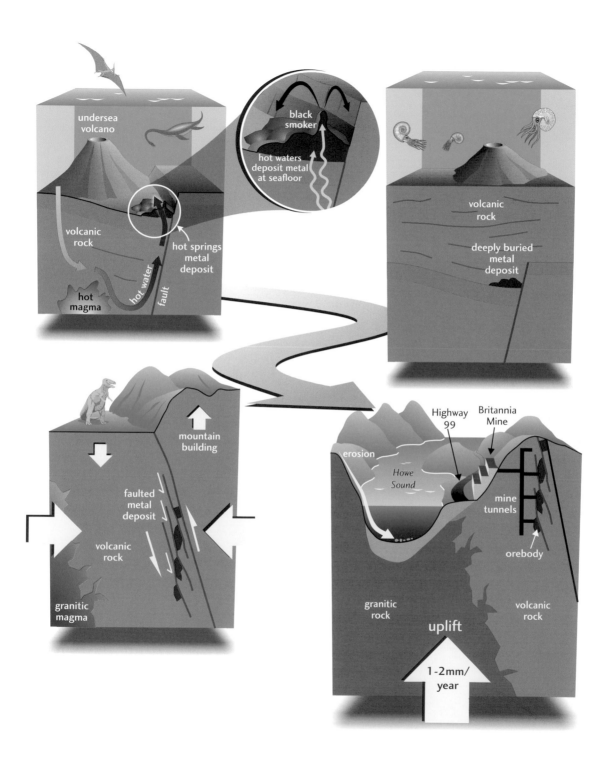

undersea volcano

volcanic rock

hot springs metal deposit

hot magma

hot water

fault

black smoker

hot waters deposit metal at seafloor

volcanic rock

deeply buried metal deposit

mountain building

faulted metal deposit

volcanic rock

granitic magma

erosion

Howe Sound

Highway 99

Britannia Mine

mine tunnels

orebody

granitic rock

volcanic rock

uplift

1-2mm/ year

Metals

Most rocks contain small amounts of metals such as copper, lead, zinc, silver, and gold. Those concentrations of metals are too low to be profitably extracted. In rare circumstances, however, Earth processes have created rocks with high concentrations of metals, far above their normal or "background" levels. These metal-rich rocks (orebodies) can be mined and the metals profitably extracted from them. Extraction can be done either underground, using tunnels to reach the orebody (as at the famous Sullivan Mine in the Kootenays) or at the surface by **open-pit mining**, which involves removing overlying metal-poor rocks to reach the orebody (as at the Brenda Mine on the Coquihalla Highway connector). Many small metal deposits, or "prospects," have been found around Vancouver and three became mines. The Northair Mine near Whistler produced silver, gold, lead, and zinc from 1977 to 1982, and the Giant Mascot Mine northwest of Hope produced nickel and copper from 1958 to 1974. But by far the largest metal mine in the area, and one of the largest in B.C., was the Britannia Mine at Britannia Beach along Howe Sound.

A "black smoker" hot spring on a submarine ridge in the eastern Pacific Ocean. The black smoke consists of fine mineral particles that precipitate when metal-rich hot water mixes with cold sea water. The hot water flows from a chimney composed of iron, copper, and zinc sulphides and is being sampled by the mechanical arm of a deep-sea submersible. The hot spring supports a unique ecosystem that includes pink tube worms that produce energy by oxidation of sulphur in the hot water. Geologists think that the Britannia orebody formed from similar hot springs on an ancient ocean floor 100 million years ago.

Coal

Coal forms from the accumulation, burial, and decay of plant matter in freshwater or brackish swamps and marshes. It is used to generate electricity and to produce steel from iron. Large amounts of coal mined in the Rocky Mountains in eastern B.C. are transported by rail across the Fraser Lowland to Deltaport on the Fraser delta, where it is loaded onto freighters and shipped to Asia.

Aerial view of Deltaport at the western front of the Fraser delta northwest of Tsawwassen. This facility exports coal from mines in the Rocky Mountains to markets in Asia.

Coal seams within Cretaceous sedimentary rocks on eastern Vancouver Island have been mined from the mid-1800s until today. Large, long-lived mines at Nanaimo exploited coal in three beds (the Douglas, Newcastle, and Wellington seams) that dip eastward beneath the city and Nanaimo Harbour. More than 50 million tonnes of coal were produced from this field. The coal seams were mined in a maze of underground tunnels that still exist, although they are no longer accessible. As many residents of Nanaimo are aware, the ground beneath them is somewhat like Swiss cheese! The only coal mine on Vancouver Island that is still operating (at least it was still open when this book was written) is near the Quinsam River, west of the town of Campbell River. The Quinsam mine, unlike those at Nanaimo, is an open-pit operation. Open-pit mining is energy-intensive and has been made possible by large trucks and excavators that were not available when the Nanaimo coal field was being exploited.

Coal has taken some lumps as a source of energy in recent years. It commonly contains sulphur, cadmium, and noncombustible mineral residues that, along with the greenhouse gas carbon dioxide, enter the atmosphere when the coal is burned. Most new thermoelectric generating plants now use oil or natural gas to produce electricity. Coal, however, is still the most widely used fuel in many developing countries, including China.

Evolving technologies may breathe new life into old "King Coal." Coal beds contains large amounts of methane, which is a very clean fuel. Methane can be extracted from coal by drilling into coal-bearing rocks and pumping water from the drill holes. When the water is removed, methane gas moves into the drill holes. Production of coal-bed methane is a much more environmentally friendly use of coal than combustion of the raw material.

Petroleum and natural gas

Western civilization is heavily dependent on petroleum and natural gas for motive power, lubrication, heating, dyes, drugs, and many synthetics. Petroleum and natural gas are complex mixtures of hydrocarbons, which are organic compounds consisting solely of hydrogen and carbon. Petroleum is refined to produce gasoline, naphtha, kerosene, lubricating oils, plastics, and many other commodities.

Petroleum and natural gas are derived from the breakdown of the remains of incompletely decayed plants and animals trapped in deeply buried sedimentary rocks. These fluids migrate from the rocks in which they form to shallower depths where they accumulate in pores, cavities, and fractures beneath impermeable rocks. Petroleum and natural gas are extracted by means of wells. Most modern wells are bored by a rotary process, in which a drilling bit is made to turn at the lowest point of the hole while mud is pumped from the surface to lubricate the cutting action and flush away the rock fragments produced by the action of the bit. The mud also creates pressure inside the well, thus supporting the sides until a casing can be inserted. Some wells must be drilled many kilometres before petroleum deposits are reached, and many wells are now drilled offshore from platforms anchored to the ocean bed. The crude oil is sent from a well to a refinery in pipelines or tanker ships.

The large quantities of petroleum and natural gas that are burned as fuels produce most of the air pollution in both industrialized and developing countries, and oil spilled from tankers and offshore wells has polluted oceans and coastlines.

Opposition of conservationists has delayed the development of many oil deposits. However, because the need for oil is so great, it may not prove feasible to hold up the exploitation of these resources until other energy sources are fully developed. British Columbians will face this issue in coming years as the debate over offshore exploration on the west coast of Canada heats up.

Life as we know it would not be possible without gasoline, which is a product of petroleum. Our dependency on this convenient fuel has led to major environmental problems, such as air pollution.

Unlike Alberta, B.C. produces no petroleum. There are, however, many producing natural gas fields in the northeastern part of the province, centred around Fort St. John and Fort Nelson. Geologists think that the rocks beneath the continental shelf east and southeast of the Queen Charlotte Islands (Hecate Strait and Queen Charlotte Sound) and possibly beneath the Strait of Georgia may contain large quantities of petroleum and natural gas. The thick Cretaceous and Tertiary sedimentary rocks in these areas are similar to rocks in other parts of the world that contain hydrocarbons. Oil seeps on the seafloor off the Queen Charlotte Islands indicate that the rocks in Hecate Strait contain at least some petroleum, but no commercially viable reserves are known at this time.

As of March 2003, there is still a moratorium on offshore oil and gas exploration in B.C. because of concerns about oil pollution. The B.C.

Geothermal power plant, Wairakei, New Zealand.

*Meager Creek hot springs. The hot springs are fed by waters that become heated beneath Mount Meager and rise to the surface along faults. Silica and calcium carbonate (lime) precipitate as the waters cool, forming crusts (**sinter**) that cement loose, porous gravel.*

government, however, is considering lifting this moratorium and allowing petroleum companies to conduct the geophysical surveys and drilling required to assess the resource. The social and economic issues are complex and deserve informed, thoughtful public discussion. Offshore oil exploration and production are not risk-free, but technology and standards have improved immensely in recent decades. Oil and gas are now routinely extracted without incident from beneath the continental shelf in many areas, for example the North Sea off Great Britain and Norway and the Grand Banks off Newfoundland and Nova Scotia.

Geothermal energy

In areas with active volcanoes and young volcanic rocks, such as Iceland, Italy, the western United States, and New Zealand, rocks are unusually hot at very shallow depths. It is possible to exploit these hot zones to produce electrical energy. Groundwater percolating down into hot rocks along fractures and faults heats up to temperatures in excess of 100° C, the boiling point of water at the

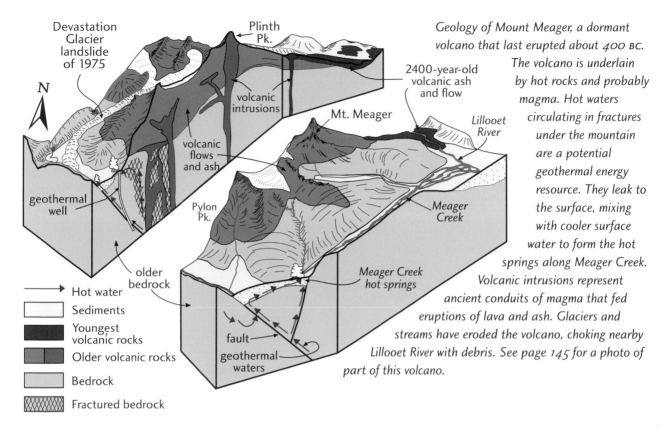

Devastation Glacier landslide of 1975

N

Plinth Pk.

volcanic intrusions

2400-year-old volcanic ash and flow

Mt. Meager

Lillooet River

volcanic flows and ash

geothermal well

Meager Creek

Pylon Pk.

older bedrock

Meager Creek hot springs

→ Hot water

Sediments

Youngest volcanic rocks

Older volcanic rocks

Bedrock

Fractured bedrock

fault

geothermal waters

Geology of Mount Meager, a dormant volcano that last erupted about 400 BC. The volcano is underlain by hot rocks and probably magma. Hot waters circulating in fractures under the mountain are a potential geothermal energy resource. They leak to the surface, mixing with cooler surface water to form the hot springs along Meager Creek. Volcanic intrusions represent ancient conduits of magma that fed eruptions of lava and ash. Glaciers and streams have eroded the volcano, choking nearby Lillooet River with debris. See page 145 for a photo of part of this volcano.

surface. Drill holes can tap these superheated fluids. As hot water rises up a hole, it boils, expands, and transforms to steam, which can then be used to turn a turbine and produce electricity.

In the 1970s, BC Hydro investigated Mount Meager as a possible **geothermal** energy source. Holes have been drilled to 3 km depth at Meager Creek, on the south flank of the volcano, to determine the geothermal power potential of the mountain. Surface waters seep down fractures under Mount Meager and encounter hot rocks at shallow depths. These waters become heated to temperatures up to 250° C. The heated waters then rise along faults and fractures, and in places reach the surface to form hot springs such as the popular Meager Creek hot springs. The hot springs at Meager Creek have temperatures as high as 59° C and are rich in sodium, silica, chloride, and bicarbonate.

Large amounts of groundwater beneath Mount Meager are hotter than 100° C, which is the boiling point of water at the ground surface. Exploratory drilling by BC Hydro indicated that it may be possible to drill holes into hot rocks and tap superheated water, producing steam that can drive turbines and thus generate electricity. BC Hydro, however, decided not to pursue the project for economic reasons, concluding that such electricity would be more expensive than that produced by burning fossil fuels or damming streams. Nevertheless, a private company recently has taken a second look at the geothermal energy potential of Mount Meager. It too drilled holes in Meager Creek valley, and it may develop the resource if it receives permits from the provincial government to proceed. An important factor in a decision by government to approve this project is the high risk of landslides in Meager Creek valley. The power production and transmission system would, of course, have to be safe from landslides through the entire life of the facility.

Energy can also be extracted from subsurface water. Shallow groundwater in the Lower Mainland is a little warmer than the land during the winter and is colder than the land during the summer. With the aid of efficient heat pumps, thermal energy extracted from this groundwater can heat homes in winter and cool them in summer.

Clay products

Clay is used to produce a variety of products including tiles, bricks, and drainage pipes. Its use has declined in recent years because of substitution of other materials, notably petrochemical products and concrete.

Clay has been mined at Haney and at Clayburn. The clay at Haney, near Maple Ridge, came from Ice Age glacial-marine sediments that are about 13,000 years old. These sediments were deposited in the sea as glaciers retreated from the Fraser Lowland (see page 51). The Haney mine ceased operation in the 1970s.

Today, there is no trace of the mine — it is a suburban neighbourhood. The clay at Clayburn, east of Abbotsford, has a very different source. It comes from Tertiary mudstone exposed at the base of Sumas Mountain. Twenty-five thousand tonnes of clay products were produced at Clayburn in 1998, and the mine is still producing today.

Sand and gravel

Sand and gravel are important construction and landscaping materials. They are used as fill, and gravel is an important component of concrete and asphalt. Without sand and gravel, we would have no paved roads, sidewalks, or high-rise buildings.

In the Fraser Lowland, this vital resource is extracted from Ice Age and modern sediments. Sand dredged from the channel of the Fraser River downstream of New Westminster is used as fill and as weight to compact highly compressible sediments on the Fraser delta prior to construction. Gravel is extracted from Ice Age sediments at numerous pits in the Coquitlam River valley and in

Gravel pit in Ice Age sediments south of Abbotsford.

Gift from a volcano

About 2400 years ago, Mount Meager blew its stack in a violent eruption somewhat similar to the 1980 Mount St. Helens eruption in Washington state. Fine tephra (ash) from the Mount Meager eruption was blown eastward across B.C. and Alberta. The eruption also blanketed the eastern flank of Mount Meager in a thick layer of much coarser tephra, consisting of particles of pumice up to boulder size. In the 1990s, an imaginative entrepreneur opened a pumice mine at the base of Mount Meager. The pumice is trucked to a nearby site where it is sorted into several size fractions for different uses. The sorted pumice is then moved to Squamish where it is stored until marketed. Pumice is used, among other things, for landscaping, to produce stone-washed blue jeans, and as an additive in some soaps.

the Fraser Lowland east of Aldergrove. Many of these mining operations will shut down in coming years due to urban encroachment and complaints about dust and noise. The gravel pits in the Coquitlam valley, for example, will be forced to close because the sites can no longer be used for industrial purposes due to suburban development on Westwood Plateau.

Alienation and exhaustion of local sources of gravel require that the resource be brought in from outside Vancouver. This, however, is not so easy. It is not feasible to move gravel more than several tens of kilometres by truck because of the high cost of this type of transportation. Transport by rail and barge are two possible solutions, and an increasing amount of the gravel used in Vancouver is now barged in from a gravel pit at Sechelt. Another solution is to use crushed rock in place of gravel. Crushed rock is used for the beds of railway tracks but is not a suitable substitute for gravel for some purposes.

Pumice mine at Mount Meager. The pumice was deposited during the eruption of Mount Meager in about 400 BC.

Parting comments

THE PUBLIC often views geology as a secondary science, less important and less interesting than physics, chemistry, or biology. In fact, some people do not consider geology a science at all, but rather an interdisciplinary field of study combining arts and science.

We passionately believe that geology is not only a science, but an absolutely essential science for today's world. It uniquely combines field and laboratory observations, and integrates physics, chemistry, and biology in its attempt to understand how the Earth works. It does this, not in an abstract way, but by making observations and measurements at actual physical locations on Earth. The complexity of the Earth is profound — there are labyrinthine interconnections among rocks, the oceans, the atmosphere, and organisms, with positive feedback cycles that are as yet poorly understood. The Earth system seems to be self-regulating, leading some to draw parallels between it and a complex living organism — the Gaia concept of James Lovelock and Lynn Margulis. No science except geology is capable of providing a full understanding of this complex system, although this understanding is possible only if geologists work hand-in-hand with other scientists.

Why is this important? There are now more than six billion people on this blue planet of ours, and our numbers are expected to climb to ten billion in the next fifty years. For the first time in the history of our species, the cumulative human impact on our planet exceeds that of all natural processes combined. We are degrading Earth's soils, oceans, fresh water, and forests; and we are contributing to the reduction in the number of plant and animal species at an alarming rate. Most people living in cities have forgotten that they are inescapably connected to the Earth that supports them. Without an understanding of how our planet works and without human behaviour that accords with this knowledge, we will reduce Earth's ecological carrying capacity and pass an impoverished planet on to our descendants. Geologists understand this more than most people; they strive to understand how natural systems function and how human activities interplay with these natural functions. We hope that this book will illustrate, in a small way, that we are indeed connected to a constantly changing Earth.

We also hope that we've achieved another of our goals — to get you interested in the wonders of geology. We are passionate about this science and hope that you've caught the bug from us as you read our book. If you want to learn more about the geology of Vancouver, B.C., or Canada, abundant in-person, print, and web resources are available to you; these are described in the next section. If you are a student, consider a career in earth science. Jobs are available in the mineral, oil, and gas industries, but earth scientists also teach in schools, work for environmental and engineering companies, universities, and government agencies at all levels. Earth scientists influence natural resource policies and land-use decisions, and they play pivotal roles in reducing the risks to people and property from earthquakes, landslides, floods, and other hazardous natural processes. Finally, a career in earth science is fun. You get to test your wits against the mysteries of the Earth. You are a professional who gets to travel and see some of the beautiful places on Earth. And you get a whole new perspective on Time.

Want more information?

THERE ARE surprisingly few non-technical books on the geology of southwestern B.C. Many of the books that do exist deal primarily with a single aspect of the geology, for example fossils or hot springs. The technical literature on B.C. is rich, however, and increasing amounts of information are becoming available on the web.

The publications listed below are a starting point for those interested in exploring the geology of the region in more detail. They are divided into two groups: non-technical books, maps, and posters of broad interest that are readily available; and technical books, reports, and other materials that contain abundant useful information but require some knowledge of geology and are somewhat harder to find. Many of these books and reports contain comprehensive bibliographies that are excellent reference sources and will lead you deeper into the fascinating science of geology. A list of relevant web sites is included after the list of technical publications.

For personal attention

A good place to start is the Geological Survey of Canada (GSC) office in Vancouver. Visit the GSC Sales Office at the corner of Robson and Seymour (address below) and check out the maps and books that are displayed and sold there. Or visit the GSC library on the 15th floor of the same building. Friendly, knowledgeable staff will help you. If they can't answer your questions, they will refer you to one of the GSC staff geologists. The address is:

Geological Survey of Canada
101 – 605 Robson Street
Vancouver, BC V6B 5J3
Phone 604-666-0529
Internet, sales office http://www.rodus.com/shop/
Internet, general office http://www.nrcan.gc.ca/gsc/pacific/vancouver

If you are interested in fossils, consider joining the Vancouver Paleontological Society. This group of amateur and professional paleontologists offers meetings, lectures, and field trips to local fossil localities. Amateurs and professionals work closely together, and many amateurs have made important contributions to the science of paleontology. Contact them at:

The Vancouver Paleontological Society
Centrepoint Post Office
Box 19653
Vancouver BC V5T 4E7
http://www.vcn.bc.ca/vanps/welcome.html

Books, maps, and posters for the general reader

These publications are non-technical and are intended for a wide audience. Most are for sale at the Geological Survey of Canada address given above. They are also available in the GSC Vancouver library, which is open to the public, (same address; phone 604-666-3812 for hours) and through many public libraries.

Armstrong, J.E. 1990. *Vancouver geology.* Geological Association of Canada, Cordilleran Section, Vancouver, B.C.
> A non-technical summary of the rocks, sediments, and geological evolution of the Lower Mainland. The book includes descriptions of sites of local geological interest.

Cannings, S. and R. 1999. *Geology of British Columbia: a journey through time.* Greystone Books, Vancouver, B.C.
> A highly readable account of the geological evolution of B.C., focusing on links between the geological history and the history of life.

Clague, J.J., Turner, R.J.W., and Shimamura, K. 1999. *Vancouver's landscape.* Geological Survey of Canada, Open File 3722 [poster, one sheet, English and French versions].
> The centrepiece of this colour poster is a detailed three-dimensional, shaded-relief map of the Lower Mainland. Photographs of representative landforms in the region and explanatory text "dance" around the map. The poster can be viewed on the web at http://geoscape.nrcan.gc.ca/.

Ludvigsen, R. and Beard, G. 1997. *West coast fossils; a guide to the ancient life of Vancouver Island.* Second edition. Harbour Publishing, Madeira Park, B.C.
A concise, thorough, readable book on the fossils of Vancouver Island written by two leading paleontologists. The book includes maps and more than 250 fossil photographs, as well as information for locating, collecting, studying, photographing, and preserving fossils.

Mathews, W.H. 1975. *Garibaldi geology: a popular guide to the geology of the Garibaldi Lake area.* Geological Association of Canada, Cordilleran Section, Vancouver, B.C.
This field guide describes the geology of the Garibaldi Lake-Black Tusk area of Garibaldi Provincial Park. It focuses on volcanoes, glaciers, and landforms.

Mustard, P.S., Hora, Z.D., and Hansen, C.D. 2003. *Geology tours of Vancouver's buildings and monuments.* Geological Association of Canada, St. John's, NL.
Five tours that focus on the geology of selected historic and modern buildings and monuments. The book describes rock types, fossils, minerals and other interesting features of the stone used in the buildings and includes comments on how these relate to the changes in building techniques and architectural styles.

Roddick, J.A. 2001. *Capsule geology of the Vancouver area and teacher's field-trip guide.* Geological Survey of Canada Open File 4022.
This field guide describes, in concise, easily understood terms, the geology of several sites of local interest, including Point Grey, Queen Elizabeth Park, Stanley Park, and Caulfeild Park.

Turner, R.J.W., Clague, J.J., and Groulx, B.J. 1996. *Geoscape Vancouver — living with our geological landscape.* Geological Survey of Canada, Open File 3309 [poster, one sheet, English and French versions].
A large-format colour poster featuring relevant earth science issues in the Vancouver area. The poster can also be viewed on the web at http://geoscape.nrcan.gc.ca/.

Turner, R.J.W., Clague, J.J., Groulx, B.J., and Journeay, J.M. 1998. *GeoMap Vancouver — geological map of the Vancouver metropolitan area.* Geological Survey of Canada, Open File 3511 [poster, one sheet, English and French versions].
This large-format geological map shows the surface geology of the Lower

Mainland on a three-dimensional, shaded-relief base. The map units are illustrated with photographs in the legend. Smaller maps focus on earthquake ground motion, earthquake liquefaction, flood hazard, landslides, and groundwater and aquifers. The map can be viewed on the web at http://geoscape.nrcan.gc.ca/.

Turner, R.J.W. and Clague, J.J. 1999. *Temperature rising — climate change in southwestern British Columbia.* Geological Survey of Canada Miscellaneous Report 67 [poster, one sheet, English and French versions].
   A large-format poster that includes impacts of climate change on landslides, floods, and sea-level rise in southwestern B.C.

Turner, R.J.W., Page, J., Klassen, M., Quo Vadis, H., and Jensen, A. 2000. *Vancouver rocks.* Geological Survey of Canada Miscellaneous Report 68 [poster, one sheet, English and French versions].
   This colour poster describes the rocks that underlie the Vancouver region using well known sites such as Stanley Park, Whistler, and Mount Baker as examples. It also contains information on the genesis of the rocks and their three-dimensional relationships. The poster can be viewed on the web at http://geoscape.nrcan.gc.ca/.

Woodsworth, G. 1999. *Hot springs of western Canada,* second edition. Gordon Soules Book Publishers, West Vancouver, B.C.
   Although primarily a field guide for the recreational user, the book has a good introduction to the geology of hot springs, water temperature and composition, and a comprehensive reference list.

Yorath, C.J. 1990. *Where terranes collide.* Orca Book Publishers, Victoria, B.C.
   The story of the evolution of the Canadian Cordillera, from Calgary to Vancouver Island. The book gives the reader a feel for how geologists actually work. A good account of plate tectonic processes as they relate to western Canada.

Yorath, C.J. and Nasmith, H.W. 1995. *The geology of southern Vancouver Island: a field guide.* Orca Book Publishers, Victoria, B.C.
   The story of the geologic evolution of Vancouver Island, with emphasis on the assembly of crustal plates, or terranes, by plate convergence and subduction off Canada's west coast.

Technical publications

These publications contain abundant useful information but require some knowledge of geology to understand. They are harder to find than the general interest books listed above, but most are available in the library and sales office of the Geological Survey of Canada in Vancouver at the address above. Most university and college libraries in B.C. have many of these publications.

Armstrong, J.E. 1980. *Surficial geology, Chilliwack (west half), British Columbia.* Geological Survey of Canada Map 1487A [map, 1:50,000 scale, one sheet].
  This colour map and the three companion maps by Armstrong (*Mission*) and Armstrong and Hicock (*Vancouver* and *New Westminster*), listed below, show the distribution of different types of surface earth materials in the Fraser Lowland from Vancouver on the west to Mission on the east. The four maps adjoin one another.

Armstrong, J.E. 1980. *Surficial geology, Mission, British Columbia.* Geological Survey of Canada Map 1485A [map, 1:50,000 scale, one sheet].

Armstrong, J.E. and Hicock, S.R. 1980. *Surficial geology, New Westminster, British Columbia.* Geological Survey of Canada Map 1484 [map, 1:50,000 scale, one sheet].

Armstrong, J.E. and Hicock, S.R. 1980. *Surficial geology, Vancouver, British Columbia.* Geological Survey of Canada Map 1486A [map, 1:50,000 scale, one sheet].

Clague, J.J. 1996. *Paleoseismology and seismic hazards, southwestern British Columbia.* Geological Survey of Canada Bulletin 494.
  A readable report summarizing the record of historic and prehistoric earthquakes in southwestern B.C., and the geologic evidence for great earthquakes in the region. The report discusses the likely damaging effects of a future large earthquake.

Clague, J.J., Luternauer, J.L., and Mosher, D.C. 1998. *Geology and natural hazards of the Fraser River delta, British Columbia.* Geological Survey of Canada Bulletin 525.
  A comprehensive overview of the geology of the Fraser River delta. It includes chapters on earthquakes and other natural hazards.

Journeay, J.M. and Monger, J.W.H. 1998. Interactive geoscience library, digital information for the Coast and Intermontane belts of southwestern B.C. Volume 1. Geological Survey of Canada Open File 3276.

A "digital library" of earth science information for the southern Coast Mountains. This CD-ROM includes photographs, a wide variety of maps, and descriptive material.

Monger, J.W.H. (Editor). 1994. *Geology and geological hazards of the Vancouver region, southwestern British Columbia.* Geological Survey of Canada Bulletin 481.

A collection of technical papers on the bedrock and surficial geology, landslides, earthquakes, volcanic hazards, and groundwater in the Lower Mainland.

Price, R.A. and Monger, J.W.H. 2000. *A transect of the southern Canadian Cordillera from Calgary to Vancouver.* Geological Association of Canada, Cordilleran Section, Vancouver, B.C.

A guide to the geology and tectonic evolution of the southern Rocky Mountains and southern British Columbia, revealed along a highway transect from Calgary to Vancouver. A version of this field trip running from Vancouver to Calgary is available as Geological Survey of Canada Open File 3902.

Roddick, J.A. 1965. *Vancouver North, Coquitlam, and Pitt Lake map-areas, British Columbia, with special emphasis on the evolution of the plutonic rocks.* Geological Survey of Canada Memoir 335.

A classic report on the rocks of the North Shore and the southern Coast Mountains. It focuses mainly on granitic rocks and their origin.

Woodsworth, G.J., Jackson, L.E., Jr., Nelson, J.L., and Ward. B.C. (Editors). 2000. *Guidebook for geological field trips in southwestern British Columbia and northern Washington.* Geological Association of Canada, Cordilleran Section, Vancouver, B.C.

Nine field trips featuring the geology of the Lower Mainland, the Fraser, Thompson, and Okanagan valleys, southern Vancouver Island, and the Cascade Range.

Web sites

Web sites tend to come and go. If one of the following addresses doesn't work, try using a good search engine such as Google to find the organization or title given after the address.

Vancouver geology and geoscience issues

http://geoscape.nrcan.gc.ca/.

Natural Resources Canada, Geological Survey of Canada. Geology and geoscience issues for Canadian communities, including Vancouver and Victoria. *Geoscape Vancouver, GeoMap Vancouver, Vancouver's Landscape, Vancouver Rocks,* and *Geoscape Victoria* are hosted on this site.

Bedrock geology and mineral deposits of British Columbia

http://www.nrcan.gc.ca/gsc/pacific/vancouver/earthsci/index_e.htm

Geological Survey of Canada, Pacific Division. Introduction to the geology of the Canadian Cordillera.

http://www.em.gov.bc.ca/Mining/Geolsurv/MapPlace/Default.htm

B.C. Geological Survey Map Place. Geological maps, mineral deposits, and good links to other sources.

http://www.em.gov.bc.ca/Mining/Geolsurv/PublicEducation/default.htm

B.C. Geological Survey. Background material and current information on mining and mineral deposits in B.C.

Climate and climate change

http://adaptation.nrcan.gc.ca/posters/home-accueil_en.asp

Canada Department of Natural Resources. A series of seven posters depicting the regional impacts of climate change in Canada, including southwestern B.C.

Earthquakes and tsunamis

http://www.pgc.nrcan.gc.ca/seismo/

Geological Survey of Canada, National Earthquake Hazards Program, Western Canada. Earthquake information and maps.

http://www.pgc.nrcan.gc.ca/seismo/eqinfo/links.htm

Geological Survey of Canada, National Earthquake Hazards Program, Western Canada. Links to other earthquake web sites.

http://neic.usgs.gov/
>U.S. National Earthquake Information Center, World Data Center on Seismology. Information on earthquakes world-wide.

http://www.geophys.washington.edu/SEIS/
>University of Washington. Seismology and earthquake information, with a focus on the Pacific Northwest.

http://walrus.wr.usgs.gov/tsunami/
>U.S. Geological Survey. Tsunami research and information, including animations.

http://www.geophys.washington.edu/tsunami/
>University of Washington. An interactive, on-line tsunami information resource.

http://www.pmel.noaa.gov/tsunami/tsu_links.html
>U.S. Department of Commerce, NOAA. Links to other tsunami web sites.

Landslides

http://www.em.gov.bc.ca/Mining/Geolsurv/Surficial/landslid/default.htm
>B.C. Ministry of Energy and Mines. Landslides in B.C.

Surface and groundwater

http://wlapwww.gov.bc.ca/wat/wtrhome.html
>B.C. Ministry of Water, Land and Air Protection. Ground water, aquifers, community watersheds, flood hazard management, floodplain mapping, snow surveys, air quality in B.C.

Volcanoes

http://www.nrcan.gc.ca/gsc/pacific/vancouver/volcanoes/index_e.html
>Natural Resources Canada, Geological Survey of Canada, Pacific Division. Volcanoes of Canada.

http://volcanoes.usgs.gov/
>U.S. Geological Survey. An excellent website with information on a broad range of subjects.

An introduction to "geologese," the wonderful language of geology

Active defences: Avalanche defence measures involving pre-emptive triggering of avalanches under controlled conditions, for example with explosives (see *static defences*).

Aggradation: The building up of a *floodplain* by a river or stream.

Aggrade: To build up a *floodplain* by depositing *sediment* during floods.

Andesite: A dark-coloured, fine-grained *igneous rock* rich in calcium-rich *feldspar*. Andesite contains more silicon and less iron and magnesium than *basalt*.

Aquifer: A water-bearing body of rock or *sediment* capable of providing useful amounts of groundwater to wells and springs.

Arête: A narrow, jagged, serrate mountain crest, or a narrow, rocky, sharp-edged ridge or spur, commonly present above snowline in rugged mountains sculpted by glaciers.

Artesian well: A well tapping a *confined aquifer* that contains groundwater under pressure. Water in the well rises above the level of the *water table,* but does not necessarily reach the surface.

Ash: Fine fragmental volcanic material deposited from an *eruption cloud*. The average particle size of ash is less than 4 mm.

Basalt: A black, fine-grained *igneous rock* rich in calcium *feldspar*. Basalt contains less silicon and more iron and magnesium than *andesite*. Oceanic *crust* consists mainly of basalt.

Bed: A layer of *sediment* that is relatively homogeneous and is separated from the layers above and below it by boundary planes. Beds give *sedimentary rocks* a conspicuous layered look.

Biotite: A common platy, rock-forming mineral of the mica group. Biotite is generally black or dark brown because it contains iron and magnesium. It is an important constituent of *granitic rocks* and also occurs in many *metamorphic* and *sedimentary rocks.*

Black smoker: A hot spring belching waters rich in metal-sulphide particles ("black smoke") from a chimney-like structure on the deep ocean floor. Iron-, copper-, and zinc-bearing sulphide minerals, precipitated from the hot waters, make up the chimneys and the large mounds that underlie them. Black smokers occur in active volcanic areas along ocean spreading ridges where continental plates are pulling apart.

Bomb: A volcanic rock fragment, larger than about 6 cm, that was thrown into the air from an erupting volcano and became rounded in flight while still hot.

Cenozoic: The most recent era of geologic time, extending from 65 million years ago to the present. It comprises the *Tertiary* and *Quaternary* periods.

Cirque: A deep, steep-walled recess or hollow, shaped like a half bowl and situated high on the side of a mountain, commonly at the head of a glacial valley. Cirques are produced by glacial erosion. They are common in the Coast and Cascade Mountains. Small glaciers near Whistler and Blackcomb Mountains, for example, lie within cirques.

Clay: Fine sticky *sediment* consisting of mineral and rock particles that are less than 4 microns, that is, less than *silt* size. Clay forms a pasty, plastic, mouldable, muddy mass when wet.

Coal: A combustible *sedimentary rock* containing more than 50 percent carbonaceous material that has been formed by compaction and alteration of plant remains.

Columnar jointing: Parallel, prismatic fractures in basaltic *lava flows.* The fractures form columns with hexagonal or pentagonal cross-sections and form by contraction during cooling.

Confined aquifer: An *aquifer* bounded above and below by *impermeable beds* or by beds with distinctly lower *permeability* than those of the aquifer itself.

Conglomerate: "Fossilized gravel." A *sedimentary rock* composed of pebbles, cobbles, and/or boulders set in a matrix of *sand, silt,* or *clay.*

Continental shelf: The submerged fringe of the continent extending from the shoreline to about 200 m depth. At the outer edge of the continental shelf, the seafloor slopes more steeply into the deep ocean.

Convection: The slow, lateral and vertical cycling of *mantle* material beneath the *crust,* mainly due to variations in temperature within the Earth.

Core: The innermost zone of the Earth, below about 2900 km depth. The core is 3400 km thick and lies below the *mantle.*

Cretaceous: The final period of the *Mesozoic* Era, extending from 145 million to 65 million years ago.

Crosscutting: A body of rock or *sediment* cuts across a second body and is thus younger. Crosscutting commonly results from injection of *magma* or fluidized sediment along fractures in older rocks.

Crust: The outermost layer of the Earth, which is composed of rocks rich in silicon and aluminum. The crust ranges from 8 to 70 km thick.

Crustal earthquake: An earthquake that occurs on a *fault* within the *crust.*

Dacite: A fine-grained *igneous rock* intermediate in composition between *rhyolite* and *andesite.* Dacite contains a more sodium-rich *feldspar* than andesite.

Debris avalanche: A type of landslide in which a mass of *sediment,* weathered rock, and vegetation rapidly slides down a slope.

Debris flow: A type of landslide in which a mass of water, *clay-* to boulder-size *sediment,* weathered rock, and plant debris flows rapidly down a steep valley, gully, or ravine. Debris flows are commonly triggered by heavy rain.

Delta: A low, nearly flat, triangular- or fan-shaped feature near the mouth of a stream or river. A delta is composed of *sediment* carried by a stream or river into the sea or a lake.

Dip: The amount of tilt of a surface. Technically, the angle that a surface, such as a *sediment* layer or *fault* plane, makes with the horizontal plane. Dip is measured perpendicular to the horizontal *strike* of the surface.

Distributary channel: A stream that flows away from the main stream and does not return to it, as on a *delta.*

Drainage basin: The area drained by a river or stream and all its tributaries. Adjacent drainage basins are separated by *drainage divides.*

Drainage divide: The boundary or "height of land" between two adjacent *drainage basins.*

Dyke (dike in the United States): (1) A sheet-like body of *igneous rock* that cuts across the layered, folded, or other structures of the surrounding rock (compare with *sill*). (2) An artificial wall or ridge built around a relatively flat, low-lying area to protect it from flooding.

Eocene: An epoch of the *Tertiary* Period, extending from about 55 million to 38 million years ago.

Epicentre: The point on the Earth's surface that is directly above the source or *focus* of an earthquake.

Erratic: A rock fragment carrier by a glacier or floating ice and deposited far from its source. Erratics range in size from pebbles to house-sized blocks.

Eruption cloud: A gaseous cloud of *ash* and other *pyroclastic* material created by an explosive volcanic eruption.

Esker: A long, narrow, sinuous, steep-sided ridge composed of *sand* and *gravel* deposited by a stream flowing in a tunnel at the base of or within a glacier. Eskers range in length from less than a kilometre to more than 150 km, and in height from a few metres to about 30 m.

Fault: A plane or zone within the Earth along which rocks have moved past one another.

Feldspar: A group of abundant rock-forming minerals of the general formula: $MAl(Al,Si)_3O_8$, where M is potassium, sodium, or calcium; Al is aluminum; Si is silicon; and O is oxygen. Feldspar constitutes 60 percent of the Earth's *crust* and occurs in all types of rocks, including *granites.* It ranges in colour from white to pink; some feldspar is translucent.

Fjord (or fiord): A deep, long, narrow, steep-walled inlet or arm of the sea along a mountainous coast. A fjord is the seaward end of a glacially eroded valley. Local examples are Howe Sound and Indian Arm.

Fjord lake: A lake occupying a long, narrow, steep-walled valley that has been deeply eroded by a glacier. Fjord lakes occur within and at the front of formerly glaciated mountain ranges. Harrison Lake and Pitt Lake are local examples of fjord lakes.

Flood fringe: The outer, highest part of the *floodplain,* which may be inundated during a large flood.

Floodplain: Flat, low land adjacent to a river channel. The floodplain is inundated when the river overflows its banks.

Floodway: The part of a *floodplain* kept clear of development and reserved for emergency diversion of floodwaters.

Focus (also termed **hypocentre**; see also *epicentre*): The point within the Earth that is the source of an earthquake.

Foliation: A planar fabric in a *metamorphic rock,* generally produced by the alignment of minerals under directed stress.

Fossil: The remains, trace, or imprint of a plant or animal in *sediments* or rocks of the Earth's *crust.*

Fraser Glaciation: The last major cold period of the *Pleistocene* Epoch (*Ice Age*) in British Columbia. The Fraser Glaciation began about 30,000 years ago, ended about 11,000 years ago, and was marked by the growth and decay of a large ice sheet that covered nearly all of B.C. and adjacent areas.

Geothermal: Pertaining to the heat of the interior of the Earth. Geothermal energy can be extracted from steam and hot water that occur naturally at shallow depth at some places on Earth.

Glacial-marine: A marine environment directly influenced by glaciers. The word is also used for *sediments* deposited in such an environment.

Gneiss: A banded *metamorphic rock* formed at high temperature and pressure deep in the Earth's *crust* by recrystallization of pre-existing *sedimentary, metamorphic,* or *igneous rock.* Gneiss consists of alternating dark- and light-coloured layers of different minerals.

Granite: A light-coloured, medium- to coarse-grained *igneous rock* consisting mainly of *quartz* and potassium- and sodium-rich *feldspar.* "Granite" is used loosely in this book for all coarse-grained igneous rocks containing quartz and feldspar (see *granitic rock*).

Granitic rock: A term loosely applied to any light-coloured, medium- to coarse-grained *igneous rock* consisting mainly of *quartz* and *feldspar.* Geologists discriminate many different types of granitic rocks on the basis of the types of feldspar and the amount of quartz; i.e., *granite, granodiorite,* and *tonalite.*

Granodiorite: A medium- to coarse-grained *igneous rock* consisting mainly of *quartz,* sodium-rich *feldspar,* and potassium feldspar. Granodiorite has less potassium feldspar and more sodium-rich feldspar than true *granite.*

Gravel: *Sediment* consisting mainly of rounded rock particles larger than 2 mm.

Hanging valley: A glacial valley that ends high on the steep side of a larger glacial valley. The larger valley was eroded by a trunk glacier and the smaller one by a tributary glacier. The discordance in the levels of the valley floors is due to the greater erosive power of the trunk glacier.

Holocene: The last epoch of the *Quaternary* Period, extending from 10,000 years ago to the present; roughly the period since the end of the *Ice Age.*

Hornblende: A black, dark green, or brown mineral that is common in *igneous* and some *metamorphic rocks.*

Hydrograph: A graph showing the stage (level), discharge, or velocity of a river or stream over time.

Ice Age: The period when climate was generally colder than today and ice sheets periodically covered large parts of North America, northern Europe, and Eurasia. The Ice Age is roughly synonymous with the *Pleistocene* Epoch.

Igneous rock: Rock formed from molten or partly molten material (*magma*).

Impermeable: The condition of a rock or *sediment* that renders it incapable of transmitting fluids under pressure.

Joint: A fracture in a rock. Unlike *faults,* joints do not displace the rocks in which they occur.

Lahar: A *debris flow* on a volcano. Lahars commonly happen during eruptions and during heavy rainstorms at other times.

Lava: *Magma* extruded onto the Earth's surface; also the rock that solidifies from it.

Lava dome: A dome-shaped mountain formed by the extrusion of *lava* of high viscosity.

Lava flow: An outpouring of molten *lava* from a vent or fissure in the Earth.

Levee: As used in this book, a levee is the elevated lateral margin of a *lava flow;* commonly a ridge of *basalt* or *andesite*. A levee forms when lava at the edge of the flow cools and solidifies as the centre of the flow continues to move.

Limestone: A *sedimentary rock* consisting mainly of calcium carbonate. Limestone is formed by organic and inorganic processes; some contains abundant *fossils* and represents ancient shell banks or coral reefs.

Liquefaction: The transformation of saturated granular *sediment* into a fluid by an external force, commonly an earthquake.

Little Ice Age: The period from about AD 1200 until the late 1800s, when mountain glaciers throughout the world were larger than today. The Little Ice Age was a time of variable but generally cool climate.

Love wave: A *seismic wave* that travels at the Earth's surface with motion transverse to the direction of propagation. See page 117.

Magma: Molten rock within the Earth, from which *igneous rocks* are derived.

Mantle: The zone of the Earth below the *crust* and above the *core*. It extends from 8–70 km depth to about 2900 km depth.

Mesozoic: An era of geologic time, extending from about 253 million to 65 million years ago. The Mesozoic is sometimes referred to as the Age of Reptiles.

Metamorphic rock: Rocks that have formed from pre-existing rocks by changes in mineral content and structure under high temperature, pressure, and stress within the Earth's *crust*.

Moment magnitude: A logarithmic measure of the amount of energy released by an earthquake. See *Richter scale*.

Moraine: A ridge consisting of *sediment* of *clay*- to boulder-size deposited at the margin of a glacier.

Mould: An impression made by a *fossil* in *sediment* or rock.

Mudstone: A fine-grained, commonly grey to black *sedimentary rock* formed by compression and cementation of *clay* and *silt*. Mudstone differs from *shale* in not having thin layers along which the rock easily breaks.

Nunatak: A hill or mountain completely surrounded by a glacier.

Open-pit mining: A method of mining in which an *orebody* is exploited by removing the overlying rock, as opposed to tunnelling.

Orebody: A body of rock containing valuable minerals in sufficient quantities to make mining commercially feasible.

Outwash: *Sand* and *gravel* deposited by streams in front of or beyond the margin of a glacier.

Pacific Ring of Fire: The zone of abundant earthquakes and active volcanoes that roughly follows the margin of the Pacific Ocean. It marks boundaries between several large tectonic *plates*.

Paleocene: The first epoch of the *Tertiary* Period, from about 65 million to 55 million years ago.

Peat: Material composed of partly or completely decomposed plant remains with a high moisture content. When peat is buried and heated, it changes into *coal*.

Permeability: The capacity of a *porous* rock or *sediment* to transmit a fluid, generally water.

Physiographic: Pertaining to physiography, the physical character of a landscape — its mountains, valleys, lowlands, and other features.

Plate (or **tectonic plate**): One of the large, nearly rigid fragments that form the *crust* and upper *mantle* of the Earth. Plates are 8 to 250 km thick.

Plate tectonics: A generally accepted theory based on the hypothesis that a small number (10–25) of *plates* "float" on the plastic upper *mantle* and move more or less independently of one another. Plates grind against each other like ice floes in a river. Much of the dynamic activity is concentrated at the periphery of plates, which are propelled from the rear by *seafloor spreading* and pulled from the front by *subduction*. Continents are part of plates and move with them, like logs frozen in ice floes.

Pleistocene: An epoch of the *Quaternary* Period, synonymous with the *Ice Age*. The Pleistocene follows the *Tertiary* Period and spans the interval from about 2 million years ago until the beginning of the *Holocene* Epoch, 10,000 years ago.

Pore space: The open spaces, or interstices, in a rock or *sediment*.

Porosity: The percentage of the volume of a rock or *sediment* that is occupied by interstices that may be filled with water, oil, or gas.

Porous: Having numerous interstices, whether connected or isolated.

Potassium-argon dating: A method of determining the age of a mineral or rock based on the known, natural radioactive decay of an isotope of potassium to an isotope of argon.

Primary wave (also called **P wave**): A *seismic wave* that travels through the Earth by alternating compression and expansion of material in the direction of motion. See page 117.

Pumice: A light-coloured, glassy *volcanic rock* containing abundant trapped air bubbles. Pumice is sufficiently light to float on water.

Pyroclastic: Pertaining to fragmented rock material produced by a volcanic explosion (see *tephra*).

Pyroclastic flow: A rapidly flowing mixture of *pyroclastic* debris and hot gas ejected explosively from a volcano.

Pyroclastic surge: A low-density *pyroclastic flow*. It contains more gas, is more mobile, and is less ground-hugging than a *pyroclastic flow*.

Quartz: Crystalline silica (SiO_2), an important rock-forming mineral. Quartz is the second most common mineral in the Earth's *crust*, after *feldspar*. It occurs in colourless and transparent hexagonal crystals and in crystalline masses, and is common in many *igneous, metamorphic,* and *sedimentary rocks*.

Quaternary: The period of Earth time between the *Tertiary* Period and the present, approximately the last 2 million years. It comprises the *Pleistocene* and *Holocene* epochs.

Rayleigh wave: A *seismic wave* that travels at the Earth's surface with a retrograde, elliptical motion. See page 117.

Rhyolite: A light-coloured, fine-grained *igneous rock* rich in *quartz* and sodium-rich *feldspar*. Rhyolite contains more silicon and less iron and magnesium than *andesite* and *dacite*.

Richter scale: A range of numerical values of earthquake magnitude, determined from trace deflections on a standard *seismograph* at a distance of 100 km from the *epicentre*. The Richter scale was devised in 1935 by the *seismologist* C.F. Richter. The scale is logarithmic — an increase in Richter magnitude of 1, for example from 3 to 4, corresponds to an approximate 30 times increase in energy. The threshold of significant earthquake damage is about magnitude 4 or 5. The largest earthquakes have Richter magnitudes of about 9. See also *moment magnitude*.

Rockfall: A type of landslide involving the rapid fall of a rock mass from a cliff or very steep slope. The rock mass breaks up into many blocks as it bounds down the slope.

Rock flour: Fine, unweathered *silt*- and *clay*-sized particles produced by glacial abrasion.

Rockslide: A type of landslide involving the rapid sliding of a rock mass on a slope. The rock mass generally breaks up into many blocks or fragments as it moves down the slope.

Sand: *Sediment* consisting of mineral and rock particles that are largely between 0.0625 mm (62.5 microns) and 2 mm in size. Sand is coarser than *silt* and finer than *gravel*.

Sandstone: A *sedimentary rock* consisting dominantly of *sand*-size mineral and rock particles.

Sand volcano: A low mound of *sand* produced by the expulsion of sand-laden water onto the Earth's surface. Sand volcanoes most commonly form during earthquakes.

Scaling: The precautionary removal of unstable, jointed and fractured rock from a steep slope to prevent *rockfall*.

Seafloor spreading: The process whereby oceanic *crust* is produced by convective upwelling of *magma* along midoceanic ridges. Oceanic crust moves away from the ridges at rates of 1 to 10 cm per year. See also *plate tectonics*.

Seamount: A submarine volcano or group of volcanoes rising high above the deep seafloor.

Sea stack: A small, isolated, pillar-like rocky island near a steep rocky shore. Sea stacks become detached from headlands by wave erosion. Siwash Rock is an example.

Secondary wave (also called **S wave**): A *seismic wave* that travels through the Earth by a shearing motion, so that there is oscillation perpendicular to the direction of propagation. See page 117.

Sediment: Loose, granular Earth material produced by weathering of rocks. Sediment is transported by and deposited from air, water, or ice. It may also form by chemical precipitation from solution. Sediment forms at the Earth's surface at ordinary temperatures.

Sedimentary rock: Rock formed at or near the Earth's surface at relatively low temperature and pressure by (1) compaction and cementation of rock and mineral fragments ("clastic sedimentary rock"), (2) precipitation of minerals from solution ("chemical sedimentary rock"), or (3) accumulation and alteration of plant and animal material ("organic sedimentary rock"). *Mudstone* and *sandstone* are examples of clastic sedimentary rocks. Rock salt and some *limestone* are chemical sedimentary rocks. *Coal* is an organic sedimentary rock.

Seismic amplification: An increase in the severity, or intensity, of earthquake shaking due to a change in topography or the character or thickness of the material through which *seismic waves* travel. The intensity of shaking can vary by a factor of three or more over small areas because of this effect.

Seismic waves: solid waves produced by the rupture of rock along a *fault* during an earthquake. Seismic waves move from the source to the Earth's surface, where they may damage buildings and other structures.

Seismogram: The record made by a *seismograph.*

Seismograph: An instrument that records vibrations of the Earth, especially earthquakes.

Seismologist: A scientist who studies earthquakes or the structure of the Earth's interior from natural and artificially generated *seismic waves.*

Shale: A fine-grained, commonly black *sedimentary rock* formed by compression and cementation of *clay* and *silt,* and characterized by thin layers along which the rock readily breaks. See also *mudstone.*

Shield volcano: A large volcano shaped like a flattened dome, built by flows of basaltic *lava.*

Shotcrete (also termed **gunite**): A mixture of cement, *sand,* and water sprayed on steep rock slopes along highways to prevent or reduce *rockfalls.*

Sill: A sheet-like body of *igneous rock* that is parallel to the layered structure of the rock into which it was injected (compare with *dyke*).

Silt: *Sediment* consisting of mineral and rock particles that are largely between 4 and 62 microns in size. Silt is coarser than *clay* and finer than *sand.*

Sinter: A hard crust or deposit of silica or carbonate precipitated from springs, lakes, or streams.

Spit: A low tongue of *sand* and *gravel* attached at one end to a shoreline and terminating at the other end in open water, generally the sea.

Static defences: Artificial structures that stop, slow, or deflect avalanches; for example dams, basins, mounds, and snowsheds (see *active defences*).

Stratovolcano: A generally cone-shaped volcano built of alternative layers of *lava* and *pyroclastic* debris. Mount Baker is a local example. See *shield volcano*.

Strike: The direction of the line defined by the intersection of a plane such as a *sediment* layer or *fault* with a horizontal surface.

Strike-slip fault: A *fault* along which rocks have moved laterally or horizontally past one another. See also *transform fault*.

Subcrustal earthquake: An earthquake with a *focus* in the Juan de Fuca Plate below the North American Plate.

Subduction: The process by which one crustal *plate* descends beneath another.

Subduction earthquake: An earthquake caused by the sudden slippage of one crustal *plate* over another at a *subduction zone*. Subduction earthquakes can be very large, up to magnitude 9.

Subduction zone: An elongate, narrow part of the Earth's *crust* where one *plate* descends beneath another.

Tephra: Fragmented rock produced by an explosive volcanic eruption (see *pyroclastic*). The rock fragments are ejected into the atmosphere, whereupon they fall back to the ground. Tephra is a general term applied to *sediment* ranging in size from *ash* to blocks many metres in diameter.

Terrane: A distinctive, *fault*-bounded body of rocks, thought to have been accreted, or added, to a continent at some time in the geological past (see *plate tectonics* and *seafloor spreading*).

Tertiary: The period of Earth time from about 65 million to 2 million years ago, immediately preceding the *Quaternary* Period. The first period of the *Cenozoic* Era.

Till: A heterogeneous mixture of *clay, silt, sand, gravel,* and boulders deposited directly by a glacier.

Tonalite: A medium- to coarse-grained *igneous rock* consisting mainly of calcium-rich *feldspar* and *quartz*.

Transform fault: A *strike-slip fault* along which a midoceanic ridge forming the boundary between two *plates* is offset.

Trimline: A sharp boundary delimiting the maximum extent of a glacier that has subsequently receded. It represents a sharp change in the age, composition, or density of vegetation, or the upper limit of unweathered rock on a valley wall.

Tsunami: A series of waves produced by a major disturbance of the ocean floor, most commonly a submarine earthquake, but also a landslide, volcanic eruption, or meteorite or asteroid impact. Incorrectly referred to as a tidal wave — tsunamis are not tidal.

Unconfined aquifer: An *aquifer* containing water that is not confined by *impermeable* rocks or *sediments*. An unconfined aquifer has a free *water table*.

Volcanic rock: A finely crystalline or glassy *igneous rock* that crystallizes from *magma* on the Earth's surface. It is either ejected explosively during a volcanic eruption or extruded as *lava*.

Water table: The underground surface separating fully water-saturated rock or *sediment* from overlying partly saturated or water-free rock or sediment.

Photo and figure credits

The following people and institutions supplied illustrations. Numbers refer to pages.

Brian F. Atwater 121
Britannia Beach Historical Society 76
British Columbia Ministry of Environment, Lands and
    Parks 67, 70 (top)
Graham Beard 31, 32
Dave Buchar 72
John J. Clague 8 (all except top), 12, 22 (right), 24, 35,
    38, 42, 44, 45, 46, 47, 49, 50, 57, 59, 78, 95 (bottom),
    100, 106, 110, 112, 125, 136 (bottom), 144, 145, 150, 151,
    153, 157, 159, 160, 163, 164
Alexis Clague 191
Tom Cockburn 58
George Diack 92 (top)
Elsevier Science Ltd. 131 (modified from *Quaternary
    Science Review,* v. 19, Clague et al., "A review of
    geological records of large tsunamis at Vancouver
    Island, British Columbia, and implications for
    hazards")
Steve Evans 147
Galiano Conservancy front cover, 64
Bertrand Groulx 18, 77
Al Harvey/Slidefarm frontispiece
Catherine Hickson 136 (top)
Brian Kent 102
Robert Kung 62 (top), 143
Gary Larson 132

Don Lister 105
Brian Menounos 68
Natural Resources Canada, Geological Survey of
    Canada 11, 13, 18, 20, 21, 54, 71, 77, 81, 82, 94, 103, 111,
    127 (bottom), 161
Austin Post 52, 148
Province of British Columbia 41, 73
Saskatchewan Industry and Resources 152
Kenji Satake 122
Wayne Savigny 92 (bottom)
Robert J.W. Turner 8 (top), 12, 25, 27, 37, 70 (bottom),
    75, 83, 146
Tim Turner 191
U.S. Department of Commerce, National Oceanic and
    Atmospheric Administration 130
U.S. Geological Survey 118, 138, 142
University of British Columbia 108
Vancouver Province 114
Vancouver Public Library 98
Vancouver Sun 72, 92 (top), 102
Karen L. Von Damm 156
Margaret Westwood 28 (bottom)
M.Y. Williams 108
Glenn Woodsworth 23
Duncan Wyllie 93

Acknowledgments

THIS BOOK is a labour of love, the result of five years of research and thought about earth science issues that are important to residents of the Lower Mainland. We are grateful for the support provided by our employers, the Geological Survey of Canada and Simon Fraser University, without which the book could not have been written. In particular, we thank Sandy Colvine, Paul Egginton, Cathie Hickson, and Jean-Serge Vincent for giving us "free rein" to work on this project. Generous financial support from the Geological Survey of Canada and the Canadian Geological Foundation are gratefully acknowledged.

Our ideas were forged in discussions with numerous colleagues, including Bertrand Groulx, Cathie Hickson, Lionel Jackson, Murray Journeay, Jim Monger, Peter Mustard, Jim Roddick, and Bert Struik; and were tested by educators, notably Beth Dye, Linda Gagno, Rosemary Knight, and Tim Turner. Richard Franklin, Bertrand Groulx, Kaz Shimamura, and Tonia Williams prepared the drawings, and Robert Kung produced the computer-generated landscape views. Jennifer Getsinger and Peter Mustard provided helpful reviews of drafts of the book. Special thanks to Glenn and Joy Woodsworth, who as editors and publisher greatly improved the book through their professionalism, knowledge, and affection for the subject matter.

Index

Many terms used in this book are defined in the glossary, beginning on page 175. The glossary pages themselves are not included in this index.

JOHN CLAGUE is Shrum Chair of Science at Simon Fraser University and emeritus scientist, Geological Survey of Canada. John and his graduate students are currently conducting research on natural hazards (earthquakes, tsunamis, floods, and landslides) and late Holocene climate change in western Canada. He has published over 200 professional papers on these and other earth science topics. John's other principal professional interest is increasing public awareness of science by making relevant geoscience information available to students, teachers, and the general public.

John is a Fellow of the Royal Society of Canada and President of the Geological Association of Canada. He is the recent recipient of the Royal Society of Canada's Bancroft Award and the Association of Professional Engineers and Geoscientists of British Columbia's Innovation Award for Editorial Excellence. He lives in West Vancouver with his wife Alexis.

BOB TURNER is a geologist with the Geological Survey of Canada in Vancouver. He spent his early years with the Geological Survey of Canada conducting research on ore deposits in the Yukon and southeastern B.C. In the mid-1990s, Bob turned his attention to earth science education. Together with John Clague and others, he produced a series of innovative posters, maps, and other material that explain the geology and geological issues in the Vancouver area, and the potential impacts of climate change in southwestern British Columbia. Their "Geoscape Vancouver" products have received acclaim from educators and students. Bob and John have expanded the Geoscape initiative to communities across Canada. In 2002, Bob was awarded the Neale Medal of the Geological Association of Canada in recognition of his contributions to earth science education. Bob lives on Bowen Island near Vancouver with his wife Rosemary.

RICHARD FRANKLIN is a Victoria illustrator and graphic artist. His work has been widely published, and he has won numerous awards for both his scientific illustrations and exhibited art works.